# Subsea Control and Data Acquisition

Collected papers from an International Conference 'Subsea Control and Data Acquisition' held at the City Conference Centre, London 25–26 April 1996.

The Conference was organized by the Institute of Measurement and Control, in collaboration with The Society for Underwater Technology, and was co-sponsored by

The British Computer Society
The Institution of Electrical Engineers
The Institution of Chemical Engineers
The Institution of Mechanical Engineers
The Institute of Electronic and Electrical Incorporated Engineers
The Institute of Physics
The Institution of Electrical and Electronic Engineers
Offshore and Energy
Noroil
Euroil
FFU

## The Conference Planning Committee

| | |
|---|---|
| Leslie Adriaansen | Det Norske Veritas (Chairman) |
| Clive Rees | BP International Limited |
| Chris Curran | Thetis Controls Limited |
| Dave Wollam | Kvaerner FSSL Limited |
| Ray Phillips | ABB Seatec |
| Sigurd Moe | KOS |
| Jim Cattanach | CoFlexip Stena Offshore |

# Subsea Control and Data Acquisition

Edited by:

L Adriaansen – Det Norske Veritas
R Phillips – ABB Seatec
C Rees – BP International
J Cattanach – CoFlexip Stena Offshore

Sponsored by:

Kongsberg Offshore as
Det Norske Veritas
ABB Seatec
and
FSSL

Mechanical Engineering Publications Limited
London and Bury St Edmunds, UK

First published 1996



ISBN 0 85298 993 8

A CIP catalogue record for this book is available from the British Library.

Printed by The Ipswich Book Company, Suffolk, UK

# Preface

This volume contains refereed papers accepted for the Conference on Subsea Control and Data Acquisition which was held at the City Conference Centre, London, on 25-26 April 1996. The Conference was organized by the Institute of Measurement and Control in association with The Society for Underwater Technology.

The conference programme was targeted at those in the offshore oil and gas industry who have an involvement in, or special responsibility for, subsea production control and monitoring systems as well as those in industry who are dealing with related technologies. The call for papers stimulated an excellent response of distinctive submissions on a wide range of themes associated with subsea control and data acquisition. Over the two days, four sessions were organized covering a wide range of subjects and providing a glimpse behind the theory of some of the technology currently in use and of that being developed.

Special thanks are due to the members of the Organizing Committee, who are listed elsewhere in this volume, for their assistance in dealing with the administrative load, for refereeing the papers presented herein, and for chairing sessions at the conference Equally, thanks are extended to the contributors and delegates for their support and high quality contributions.

We would also like to thank and acknowledge Kongsberg Offshore as, Det Norske Veritas, ABB Seatec, and FSSL for their generous sponsorship of this event, without which the event would not be commemorated by this hardback volume.

A special mention must go to Mr Leslie Adriaansen whose patience and hard work have brought the conference together as a coherent event.

# Related Titles of Interest

| | | |
|---|---|---|
| *Hydrocyclones 96* | D Claxton, L Svarovsky, and M Thew | ISBN: 1 86058 032 7 |
| *Fluid Machinery* | IMechE Conference 1996–4 | ISBN: 0 85298 994 6 |
| *Pressure Systems: Operation and Risk Management* | IMechE Conference 1995–4 | ISBN: 0 85298 951 2 |
| *Piping Engineering and Operation* | IMechE Conference 1993–11 | ISBN: 0 85298 863 X |
| *Pressure Surges and Fluid Transients in Pipelines and Open Channels* | BHRG 19 | ISBN: 0 85298 991 1 |
| *Pipeline Protection* | BHRG 16 | ISBN: 0 85298 976 8 |
| *Loss Prevention in the Oil and Gas Industry* | BHRG 15 | ISBN: 0 85298 975 X |
| *Hydraulic Modelling* | BHRG 11 | ISBN: 0 85298 922 9 |

For the full range of titles published by MEP contact:

Sales Department
Mechanical Engineering Publications Limited
Northgate Avenue
Bury St Edmunds
Suffolk
IP32 6BW
England

Tel: 01284 763277
Fax: 01284 704006

# Contents

## Theory and Application



## Control Fluids and Seals



## System Integration



## Service Experience

## Acoustic Technology



## High Integrity Pipeline Protection Systems



## Extending the Technology

## Theory and Application

# Advanced control and monitoring of subsea production systems

G DALEY
European Gas Turbines Limited

*Abstract*

This paper presents the results of a study into the use of advanced control and monitoring techniques in subsea production systems. The application area covered is automatic choke control, however, this has many features in common with other subsea systems and the solution techniques proposed are widely applicable. The potential of the proposed approaches for optimal field management is demonstrated using a single phase simulation of the realistic field. In addition to control, the Model Predictive strategy used also enables a simple model of the field to be obtained which is useful for fault detection purposes.

## *1 INTRODUCTION*

Modern subsea oil and gas production can encompass many different types of control system, both electrical and hydraulic. Typical examples of these are wellhead production control systems, chemical injection systems, choke control systems, and subsea separation/boosting systems. In comparison with many modern land-based process control systems, the control techniques used tend to be very traditional and relatively unsophisticated. Furthermore, despite the increasing use of subsea instrumentation, the processing of the data obtained is mainly concerned with simple logic-based fault annunciation, for observation or intervention by the operator as appropriate. The limitations of the current approach to subsea control and monitoring tend to be offset against the obvious advantages of using low risk, field-proven technology.

Although the control hardware associated with subsea production systems has much in common with similar land-based hardware, albeit in a marinised and more robust packaging, the systems engineer is presented with some particular control problems to which traditional control methods are not best suited. With the increased power of topsides computers, and control system manufacturers now making use of subsea micro-processors, the means to implement specifically tailored control schemes can be provided at relatively low cost.

In the field, the inaccessibility of the hardware means that the system operator is currently forced to make vital decisions which can have a major impact on productivity based entirely upon the available instrumentation readings; these may not yield a complete picture of the system status, and may themselves be faulty giving a false picture. With the requirement for safety of both personnel and the environment, and the constant pressure to maintain production, there are now clear incentives for implementing an integrated policy on control and monitoring which is specifically geared to improving system performance, efficiency, reliability and availability.

This paper describes the results of a study to assess the advantages of using advanced control and monitoring strategies in subsea production systems. The efficacy of various control and monitoring approaches was investigated by making use of a realistic simulation of a subsea production control scenario.

## *2 APPLICATION AREAS*

There are several areas of subsea production which could benefit from the adoption of a more sophisticated control and monitoring strategy. The areas reviewed here reflect the concerns of a number of operators whose views were obtained before progressing to a more detailed study of the potential benefits using a simulation of one specific application.

### *2.1 Common Source and Common Sink Problems*

There are two particular classes of fluid transmission line configuration commonly found in subsea systems which can cause control difficulties, namely the "common sink" and the "common source", although some systems can have elements of both.

#### *2.1.1 Flow into a Common Sink*

The common sink situation is where a number of transmission lines converge into a single line, as is the case for subsea chokes feeding a common flow line. The pressure in the flow line is a function of the summation of all of the individual choke flows. This cross-coupling of choke flows means that any change in the setting of one choke produces unwanted changes in the flow passed by the others. The entire choke control installation can be thought of as a single *multi-variable* system. Some operators report that continual adjustments are necessary to all of the field chokes as a result of such interaction. This can clearly result in increased component wear and reduced operational life. Other operators point to the need for a careful strategy when shutting in one well due to the resulting drop in manifold pressure drop which can lead to overflows from the remaining wells.

#### *2.1.2 Flow From a Common Source*

An increasingly common type of subsea system in which a number of different flow lines emanate from a larger single line or "header"is chemical injection. Nowadays, hose liner/fluid compatibility for chemical injection has become a major concern of operators, and hose liners tend to be selected for a particular duty. As the number of wells tied back to existing processing structures continues to increase, the feasibility of discrete hose injection systems on large fields becomes very questionable. One way around this problem is to use an umbilical containing a smaller number of header hoses, each dedicated to the supply of a single chemical to the subsea facilities. A subsea distribution system then feeds each well with the exact dosage required, using a remotely operated subsea control valve. In this way the hoses can be selected on the basis of single chemical usage, optimising their life from a compatibility standpoint. There is also a significant cost saving on the umbilical using this approach.

The common header umbilical, however, introduces a cross coupling between the flow rate passed by the subsea chemical injection control valves. These valves need to be proportionally controlled, and any change in the set point of one can upset the others as the pressure in the header hose changes to adjust to the new net flow demand. Clearly a control system which could automatically cater for changes in valve upstream pressure or well pressure without influencing any other field settings would be a significant improvement, both in terms of more economic use of chemicals and more precise control over the production process.

## *2.2 Complex Multi-phase Process Control*

Subsea separation and boosting systems are often required to control many dependent parameters simultaneously, usually by allowing the parameters to cycle between limits at which pre-determined control actions are invoked. Taking a typical subsea two-phase booster station as an example, pressure in a separator is used as a method of controlling production flow rate. The pressure, however, is dependent on the gas compressor speed, pump speed, gas and oil delivery line valve settings and the level of the oil surface inside the separator. As well as maintaining the oil level between the operational limits, the controller has to prevent compressor surge, and deal with system parameter changes such as temperature variation on start-up and varying oil/gas ratio.

Another example of this type is control of hydraulic downhole pumping systems. These have many of the features already discussed above in that they incorporate both the common sink and the common source problem. Moreover, in the hydraulic turbine approach the drive fluid is commingled with the production fluid leading to a nonlinearity between the system *input* and *output* chokes. The most efficient control strategy is by no means obvious and requires careful consideration of the coupled nature of the process variables.

## *2.3 Long Control Lines*

The use of long umbilicals gives rise to long time constants governing the rate of pressure wave propagation in hydraulic fluids and chemicals. The delay between control action and effect is often several minutes. In the case of chemical injection systems, for example, this results in obvious difficulties in achieving an effective flow rate control where topsides metering is used.

The problem of controlling subsea pressure with a surface choke also fits into this category. Propagation delays have been known to cause oscillatory behaviour where simple closed loop control strategies have been employed

## 2.4 Selected Application Area

In terms of potential benefits and future requirements, a more sophisticated control of subsea and topsides chokes is seen by many operators as a useful application of the technology and was therefore selected as the application area for further study. This is not only due to the coupling problem caused by a common sink but also as a result of an increasing tendency to view subsea production as a complete process. Control of the process as a whole to maximise financial returns by prolonging reservoir life, reducing water and sand production, hydrate and wax formation can then be considered. Any control and monitoring strategy however must consider reliability and maintain production in the face of sensor and actuator failure.

# 3 ADVANCED CONTROL AND MONITORING PHILOSOPHIES

## 3.1 Global Control

Based on operational requirements, three different automatic control philosophies can be considered where the aim is to regulate system variables such as pressure and flow around prescribed values. These strategies are *control from the surface choke* whereby subsea chokes are fixed at some nominal position and subsea variables are manipulated using only the surface choke; *local control* where each choke valve has a control loop that feeds back variables measured locally to the valve and *global control* in which all measured field variables are fed to a central controller which distributes control signals in a co-ordinated manner to all chokes.

Each of these strategies has specific advantages and drawbacks. For example, control from the surface choke will minimise subsea choke operations reduce wear and thereby prolong life. However, care must be taken with this strategy to correctly account for the long dead-time in the subsea response to surface changes. If this is not incorporated correctly in the design, damaging oscillatory responses can be experienced.

With a local control strategy each choke valve in the field modifies its position as a function of variables measured close to its physical location. If flow is the variable to be regulated then the choke will effectively operate as a pressure compensated flow control valve and the operator will demand a flow rate rather than a valve position. This strategy decentralises the control action and therefore, if necessary, the controller hardware and software can be also installed locally to form a *smart valve*. This approach, which is being adopted in other industries for control valve technology, also offers the facility of on-board health monitoring. Since the actuation and controlled variables are local to each other the problems of propagation delay is minimal and relatively simple feedback controllers (PID, for example) may be used. Disadvantages of the local strategy are that variable management is limited, one local loop will tend to fight the others (due to variable coupling) and constraints are not naturally handled.

A very important issue when considering the application of classical design techniques to the control of oil and gas production is the lack of a formal approach for dealing with physical constraints such as valve saturation. A classical design that behaves well without constraints can even be destabilised when these are present. Moreover, the control of the field requires the manipulation of a number of coupled variables which themselves can be the subject of constraint requirements. For example, there are often pressure-flow-temperature *no-go* regions aimed at avoiding hydrate or wax formation in the production flowline. As more is learnt about the requirements for optimum production (which may include consideration of reservoir, control equipment and pipeline life rather than only maximum continuous flow) there is a need for a flexible control strategy that is capable of achieving the optimum at any given time.

The strategies considered above have serious limitations when all of these requirements are taken into account. A more appropriate strategy is a global one where all field variables are manipulated in a co-ordinated way using the best combination of choke actions. A particularly apposite design approach based on the total requirements is Model Predictive Control or MPC. Unlike many other advanced control approaches the MPC approach was developed in industry to solve real industrial problems. There are now many different approaches to MPC but most have their roots in the dynamic matrix control of Cutler [1] which was developed to optimise the performance of refinery units and ethylene plants. Without question MPC has been the most successful and widely industrially used advanced control approach. A wide variety of applications have benefited from its application most notably in the process industries ([2-4], for example).

The MPC approach is multivariable and constraints in both actuators and controlled variables are handled naturally as part of the design process. A wide range of control actions are possible and the approach can optimise performance for a range of objectives. The approach is based on the use of simple process models which can be used to assess, offline, the feasibility of achieving a particular set of objectives.

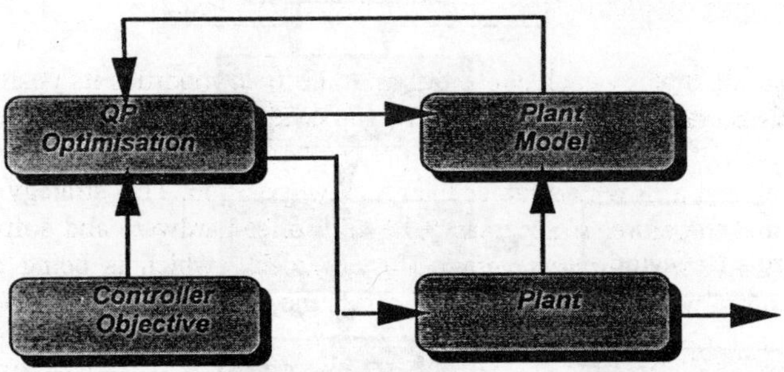


*Figure 3.1 Model Predictive Control Architecture*

The basic philosophy of the MPC approach is summarised in Figure 3.1. In essence, the plant model is used in conjunction with an optimisation algorithm (Quadratic Programming is often used) which operates in a local off-line loop until the set of control actions which come closest to achieving the control objectives are found. Once the optimum control actions have been found these are implemented on the real plant. The use of the model enables a prediction of the response several steps into the future so that the current control action can take into consideration future goals. Unlike many other optimal control approaches which require linear and quadratic cost functions, the controller objectives can be varied and can include constraints on variables, minimisation of financial costs and tracking of desired variable trajectories.

The power and flexibility of the MPC strategy is demonstrated in section 4 through its application to a simulated production system.

## *3.2 Fault Diagnosis*

In addition to a sound control approach, a high degree of fault tolerance in complex systems is an essential part of the overall system design. Fault tolerance can be achieved not only by increasing the reliability of individual plant components but also by detecting, locating and diagnosing faults at an early stage in their development. Traditionally, fault tolerance in complex engineering systems is achieved through the use of ***physical redundancy***. The essence of this approach is to locate several hardware elements in parallel in order to provide protection against localised failures. The drawbacks of this approach are the cost of purchasing and installing the extra equipment, the additional space required to accommodate it and the fact that similar components will have similar life expectancies.

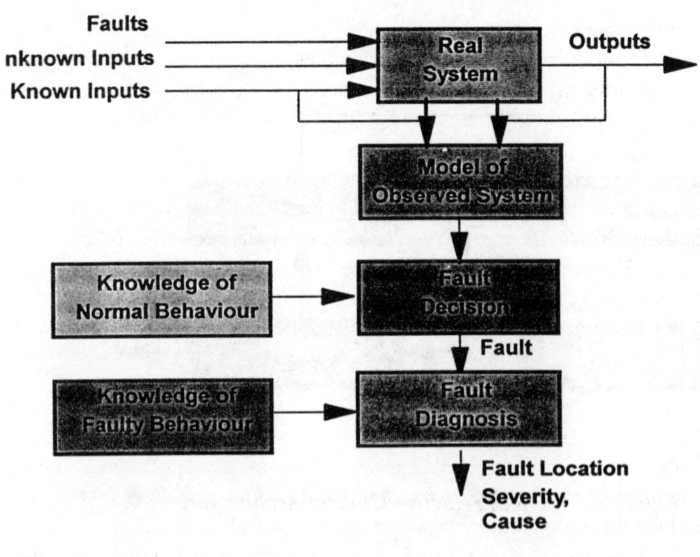


*Figure 3.2 Model-based Fault Diagnosis*

These problems, together with the advances in computer technology, have led to the development of new approaches which obviate the need for some or all of the redundant hardware. These methods, in general, operate using a more sophisticated processing of available signals and many make use of mathematical models of the system or sub-system to be monitored. The model-based approaches rely on the idea of ***analytical redundancy*** (AR) whereby the model is used in conjunction with the measurements to form a signal (commonly referred to as a residual) that has the property of being small only when the system is free of faults. When a fault develops or a failure occurs the dynamic response of the residual can be used to determine the type, magnitude and location of the fault. This process is summarised in Figure 3.2. The main AR approaches use either state estimation to provide estimates of unmeasured internal states such as pressures and flows, parameter estimation to provide estimates of physical variables such as bulk modulus and mechanical efficiency or parity relations which use system models more directly to generate a residual.

The model based methods of fault diagnosis described above have been applied to a variety of processes [5] and can readily be applied to many types of subsea system. The positive identification of faults is very important, as it is often possible to configure the control software so that alternative control or instrumentation strategies are automatically implemented. For instance, knowing that a pressure transducer is giving a false reading, it may be possible to infer what the reading should have been from a knowledge of how other system variables are behaving. This is clearly advantageous in subsea applications where the incentive to maintain production is dominant, and where work-over or ROV service is both costly and time-consuming. The AR concepts are illustrated further in the following section.

## 4 SIMULATION STUDY

In undertaking this study general systems-based design methods were adopted which concentrated on the underlying dynamic features of the plant components. This was to ensure that, although targeted initially at choke control for demonstration purposes, the methods developed would be widely applicable to the other application areas discussed in Section 2.

The purpose of using a computer model of a target system was to avoid the considerable expense and technical/commercial risk associated with a real hardware implementation. The target system needed therefore to be representative of current completion practice and to include some of the more demanding dynamic features which would best demonstrate the advantages offered by the integrated control and monitoring strategy. To ensure that the analysis performed in the study related to real life scenarios the field modelled was based on a planned field.

### 4.1 Field Layout

Essentially the field consists of eight production wells. There are two production templates, each having four production Christmas trees mounted locally. The manifold valving is such that four wells can produce through a single pipeline, as shown in Figure 4.1.

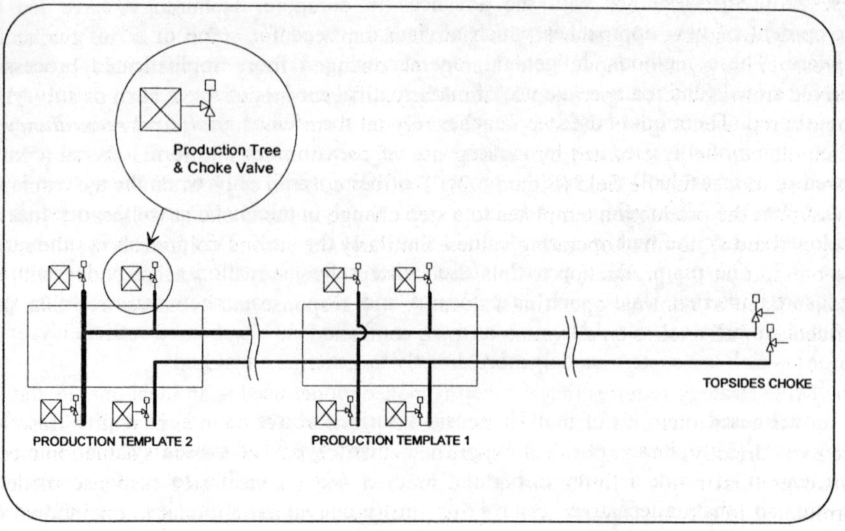


*Figure 4.1 Schematic of the Field Layout*

In order to facilitate simulation, the field was modelled as three wells at the furthest template and a single well at the infield template, as shown in Figure 4.2. Flow throughout the field model was controlled by means of a choke valve located at each production tree and one topsides.

For the purposes of this study the production fluid was assumed to be primarily liquid. This approach enabled the advantages of advanced control and monitoring techniques to be investigated without expending effort on modelling complex multiphase fluid flow. The control techniques studied are, however, believed to be equally applicable to the multiphase case. Transmision lines were modelled using a nodal description [6] so that effects on local pipeline variables could be observed.

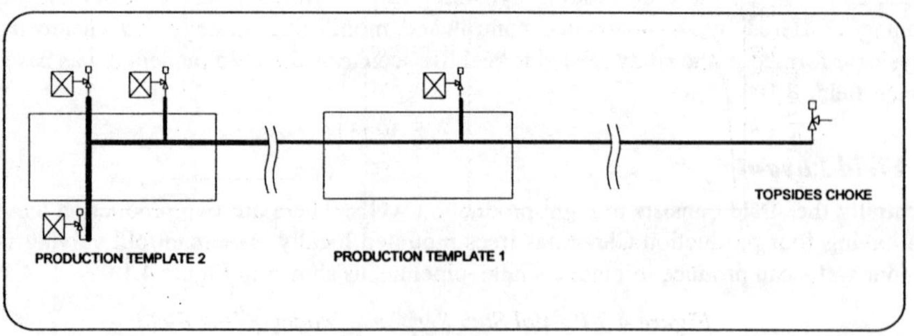


*Figure 4.2 Schematic of the Field Model Used for Analysis*

## 4.2 MPC Control

As stated in section 3.1, the key to successful application of MPC is the availability of a representative system model which may be obtained from fundamental physical laws or derived from collected operational data. Many different models have been used in MPC but a popular representation is the step response model from which the response to an arbitrary set of inputs can be determined by making use of convolution. Figure 4.3 shows a partial step response model for the field (Figure 4.2). The first column of plots shows the response of the pressure at the production templates to a step change in the choke at wellhead 1 with all other chokes fixed at nominal operating values. Similarly the second column shows the response of the pressure at the production templates to a step change in the surface choke with all other chokes fixed at nominal operating values. A step response model is required that shows the influence of all chokes on all variables to be controlled.

The MPC strategy requires that the step response model used is an ideal one, ie. the response to an instantaneous unit change in the choke position. Of course, in reality this cannot be achieved directly, due to physical limitations in the choke valve and its actuation mechanism. However this is not a limitation of the method and an ideal step response model can be formulated in a two stage process by first fitting a parametric model to operational data and then using these parameters to infer the ideal response. This general approach can be used to obtain models from a variety of normal operating data, including cases where more than one choke is operating. The system identification can also be carried out continually in order to make the MPC adapt to changing characteristics. As will be shown in section 4.3, continuous or repeated identification is also useful for fault detection and diagnosis purposes .

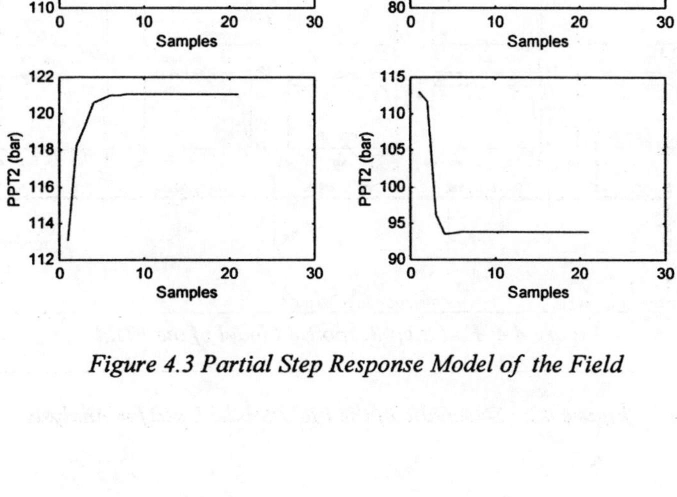


*Figure 4.3 Partial Step Response Model of the Field*

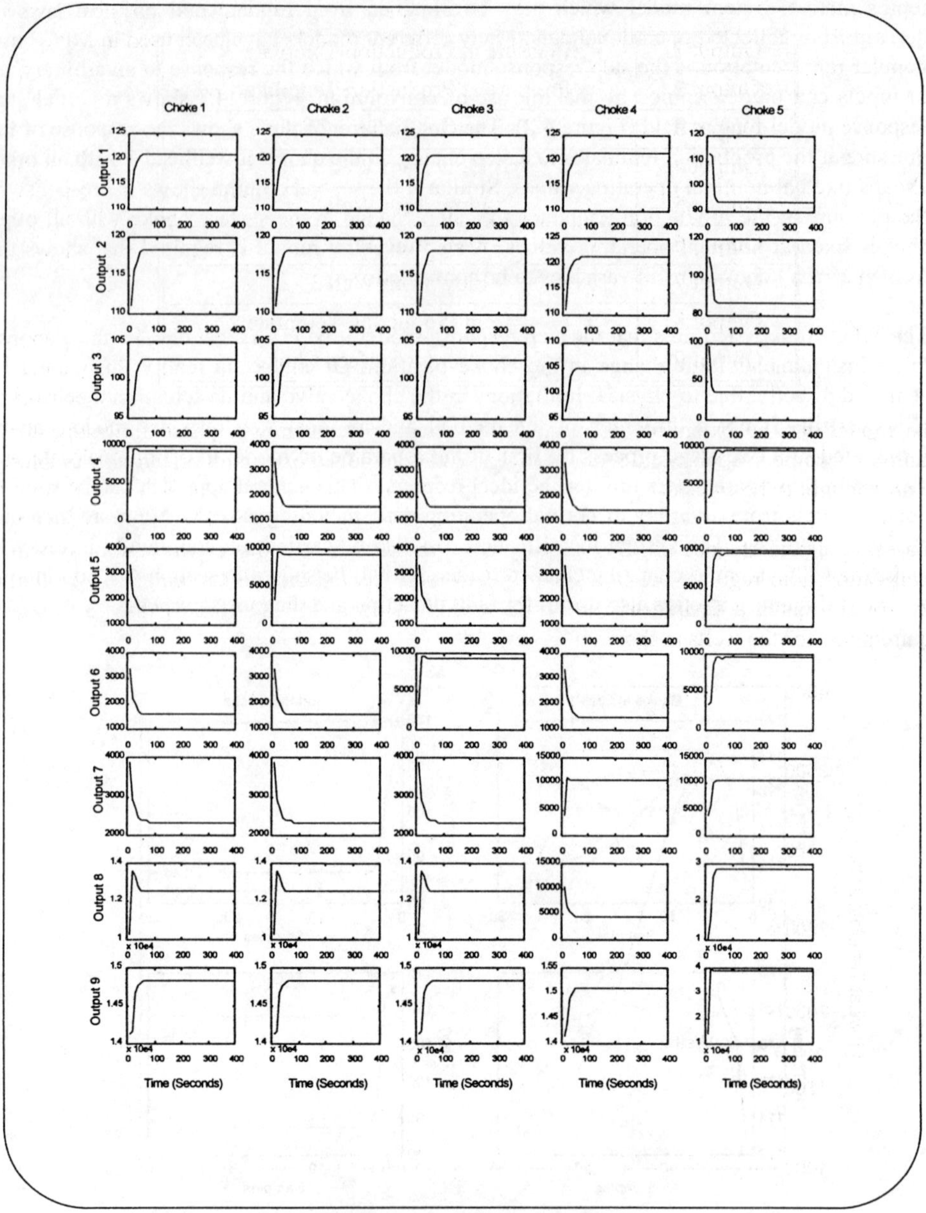


*Figure 4.4 Full Step Response Model of the Field*

Using this procedure, a full step model of the field was built up (Figure 4.4) assuming availability of the following signals (this assumption is not essential)

| | |
|---|---|
| Output 1 | Pressure at Production Template 2 |
| Output 2 | Pressure at Production Template 1 |
| Output 3 | Pressure at Surface Choke |
| Output 4 | Flow from Well 1 |
| Output 5 | Flow from Well 2 |
| Output 6 | Flow from Well 2 |
| Output 7 | Flow from Well 3 |
| Output 8 | Flow from Production Template 2 |
| Output 9 | Flow through Surface Choke |

This model can be extended if more field variables (temperature, phase flowrates, for example) become available and reduced if sensors are removed or control of certain variables are not thought to be critical.

To assess the performance of MPC a flow control scenario is considered. The control system specification is to regulate the flows out of each well feeding production template 1 so that they closely match a desired trajectory. The result of using MPC to follow three different requirements for the wells is shown in Figure 4.5.

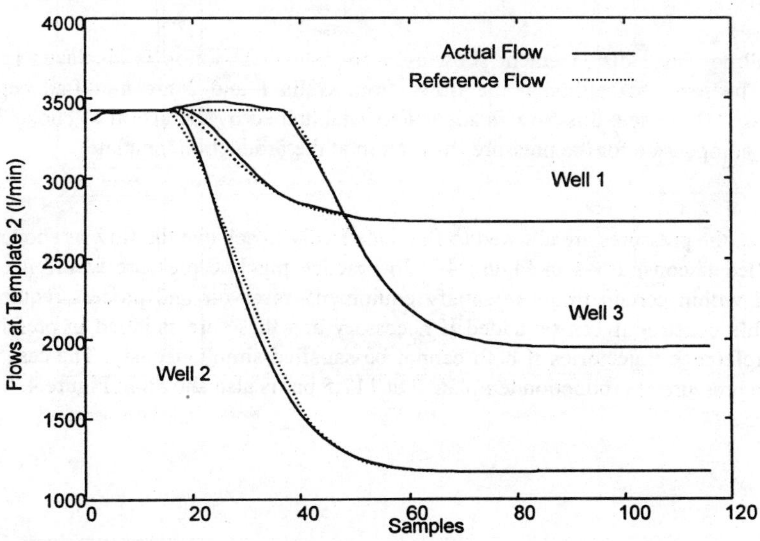


*Figure 4.5 Wellhead Production Flows : MPC Control*

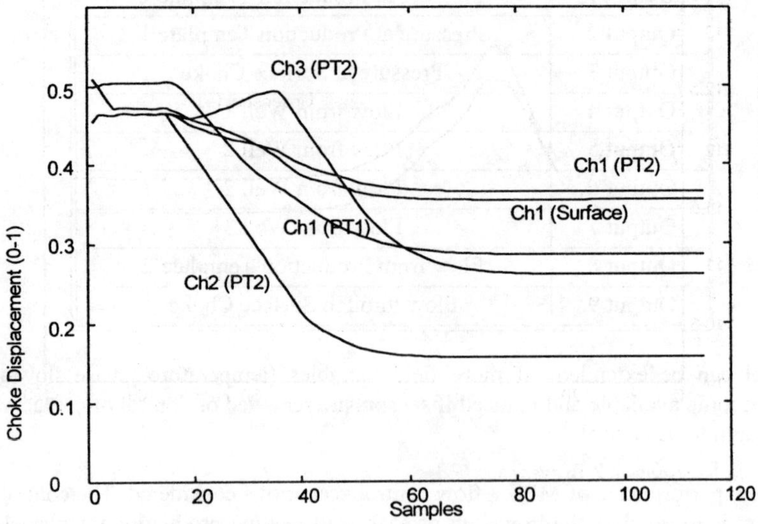


*Figure 4.6 Choke Valve Displacement : MPC Control*

The controlling choke displacement showing a co-ordinated motion is displayed in Figure 4.6. It can be seen that although the flows from wells 1 and 2 are modified before the demanded well 3 change, this flow is adequately maintained by the action of choke 3 which attempts to compensate for the pressure fluctuation at the production template.

For this case, the pressures are allowed to fluctuate freely to achieve the flow as shown by the curve labelled unconstrained in Figure 4.7. In practice pipeline pressure would need to be constrained within certain limits to satisfy equipment, reservoir and process requirements. Hard variable constraints can be added if necessary and these are satisfied in preference to achieving reference trajectories if both cannot be satisfied simultaneously. The case of hard limiting the pressure at production template 2 at 112.5 bar is also shown in Figure 4.7.

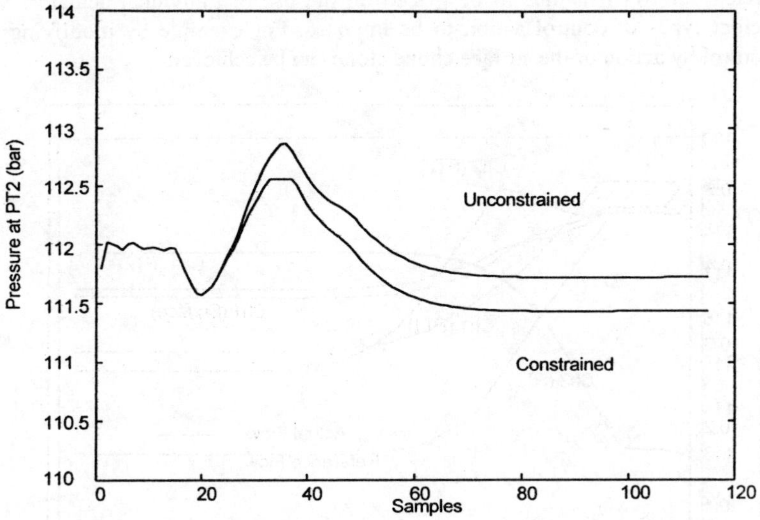


*Figure 4.7 Pressure at Production Template 2 : MPC Control*

For this limit it is also possible to achieve the demanded flows (Figure 4.8) by modifying the choke displacement.

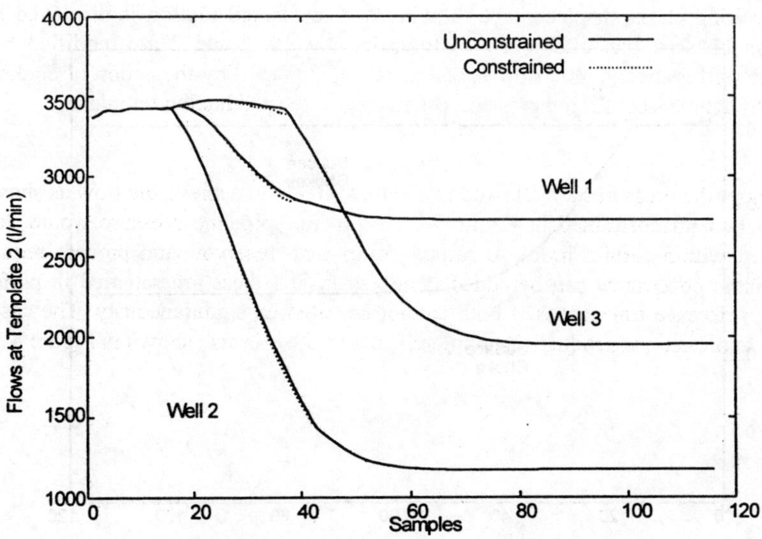


*Figure 4.8 Wellhead Production Flows : MPC Control*

The MPC strategy allows penalties to be placed on the use of individual actuators which enables particular types of control action to be imposed. For example by modifying these penalties, a control by action of the surface choke alone can be achieved.

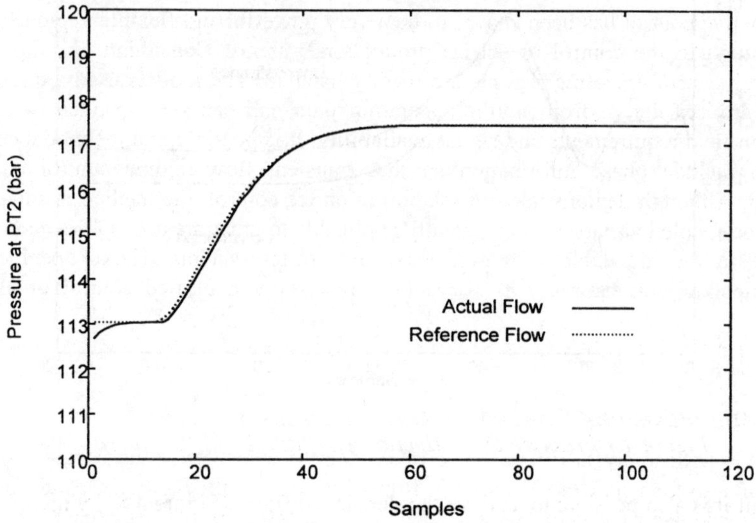


*Figure 4.9 Pressure at Production Template 2 : MPC Control*

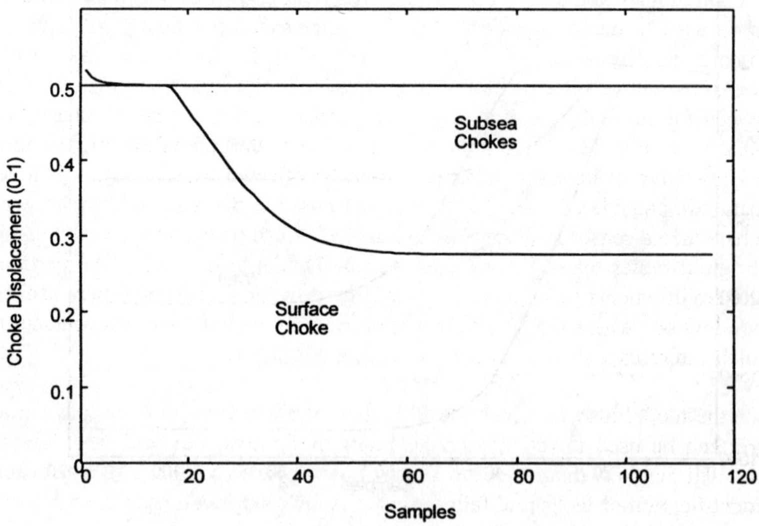


*Figure 4.10 Choke Valve Displacement : MPC Control*

Figure 4.9 shows the response of the pressure at production template 2 to an MPC strategy which has been configured for a demanded pressure increase along a reference trajectory. This trajectory can be seen to be achieved by using only the surface choke (Figure 4.10). The subsea chokes are maintained at their nominal half open position.

Model predictive control has been shown to be a very powerful and flexible approach which is ideally suited to the control of subsea production systems. Constraints in choke valve operation and system operating regions are readily handled. The models used are relatively straightforward to derive from normal operating data and can be expanded or reduced depending on field requirements and signal availability. It is possible that the model could be extended to include phase information so that transient flow regime control could be incorporated. Although demonstrated in relation to choke control, the method is suitable for any system of coupled variables and is equally applicable to other areas of subsea production. Indeed, due to the inevitable couplings between distinct systems (choke operation and chemical injection, for example) it is feasible to consider a unified control of all field variables.

## *4.3 Fault Diagnosis and Controller Reconfiguration*

The basic AR philosophy is to make use of models to characterise the normal operating behaviour of a system and use departures from this to identify faults. In a recent study, Deibert [7] made use of very detailed models to detect erosion and sedimentation in flow control valves. Here, a more general approach which makes use of the models needed for model predictive control is considered.

The step response model, Figure 4.4, can be used to *fingerprint* the system behaviour either periodically or continuously and can be determined from online measurements. Differences in the model can be used to determine specific faults. Although some faults will have subtle influences on the model difference which are difficult to identify, others will lead to obvious features. Sensor faults will manifest themselves as features in the error matrix (defined as the difference between the normal operation model and model at time of health check) which affect a particular row only. This clear fault pattern is due to the redundant information that comes from a knowledge of how the pressure should be effected by *all* chokes. If a sensor and actuator pair are considered in isolation it is not possible distinguish between actuator and sensor faults. Once a sensor fault has been identified it can be corrected for by using the normal model output rather than the measured signal. This approach would be particularly useful in hostile environments (such as encountered by downhole sensors) where sensor life expectancies are low but where the available useful life may be sufficient for a model to be determined which can replace the sensor function when this is lost.

Similarly, since the model also indicates the influence of chokes on *all* measured variables, the error matrix can be used to identify choke faults or failures. For this case, faults in a particular choke will manifest themselves as features in a specific column of the error matrix. The positive identification of faults and failures can also be used as a trigger in an active fault tolerant control strategy where a reconfiguration takes place to maintain operation as close as possible to the healthy condition. Reconfiguration requires the use of a suitable control strategy which is capable of exploiting the redundancy of control that is present in most multivariable systems. As has been shown, the MPC approach is flexible and can be readily reconfigured to provide differing control actions. It is therefore an ideal approach to provide a fault tolerant controller.

As an example, consider a case where the model based fault detection algorithm has detected a failed actuator on the choke at well 3 of production template 2. A reconfiguration is required to enable, as near as possible, the same flows as were achieved with all chokes operational (Figure 4.5). This can be achieved by placing a high penalty on the use of choke 3, which then forces the correct control action. The result of this reconfiguration is shown in Figure 4.11.

Although the match is not as close as the fully operational case, the reference flows are tracked with reasonable accuracy even from the well where the choke actuator has failed. The action of the chokes is shown in Figure 4.12 where it can be seen that the choke at well 3 maintains a constant position of half open. The flow from well 3 is obtained by manipulating the pressure at production template 2 and modifying the position of the other operational chokes to achieve the flows elsewhere.

Clearly the ability to reconfigure production in this way is very case specific, and in practice the latitude available for changing pressure will be limited. As has been shown, however, these limits can be built into an MPC strategy to cater for *allowable reconfiguration* should specific faults occur, to obtain the best possible performance within predefined limits.

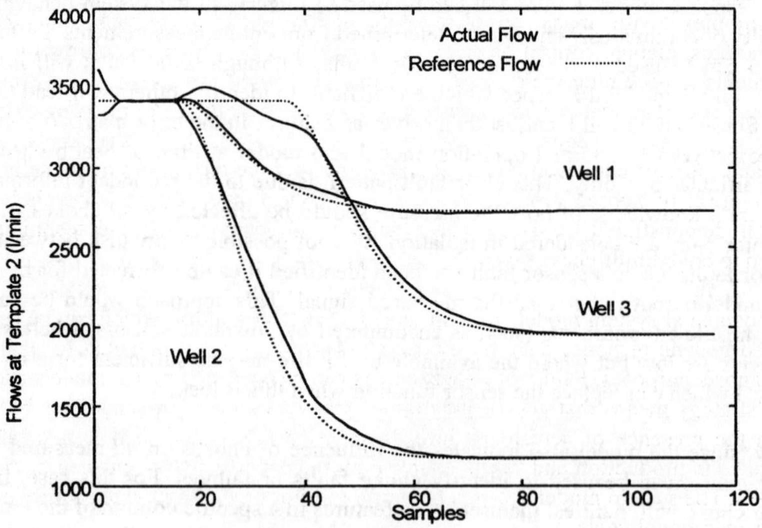


*Figure 4.11 Wellhead Production Flows : Reconfigured MPC Control*

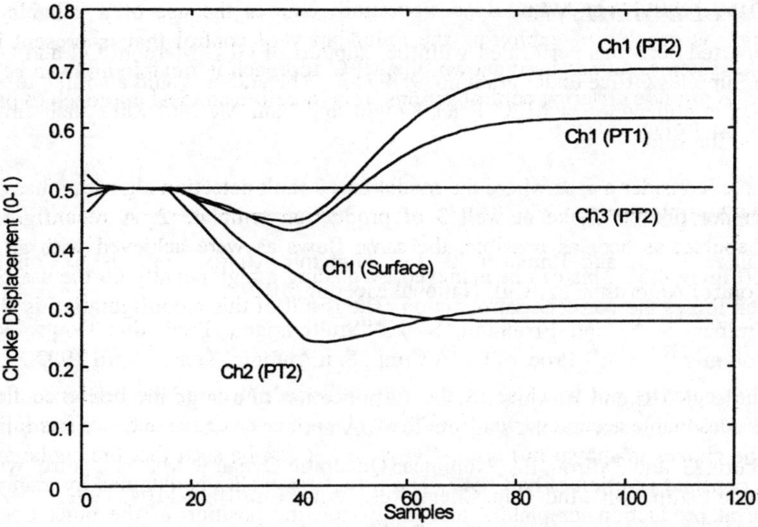


*Figure 4.12 Choke Valve Displacement : Reconfigured MPC Control*

## *5 CONCLUSIONS*

Many subsea production systems have dynamic properties that are characterised by several interdependent variables which must operate within known limits. As subsea systems become more complex, with many template and satellite wells tied back into the existing infrastructures, manual control of the many variables and sometimes conflicting operational requirements is becoming increasingly difficult. This study has shown that a strategy based around the use of model predictive control offers real benefits in terms of operability and ultimately, increased revenue. This strategy has been demonstrated in the simulated control of subsea and topsides chokes in production flowlines but the techniques are equally applicable to other areas of subsea production such as chemical injection and downhole pumping. Although demonstrated on single-phase fluid it is believed that the techniques can be extended to cover multiphase flows.

The fundamental system models used in model predictive control, which can be derived from measured general response data, have been shown to have great potential for fault detection and diagnosis purposes. Faulty or failed sensors can be corrected by the models and a flexible control strategy means that good regulation of many variables can sometimes be maintained even in the presence of actuator failure. In practice this would lead to a greatly reduced disturbance in production and would enable a more cost effective maintenance strategy to be employed. The system model is also a powerful tool for predicting the feasibility of achieving specific operational goals.

## 6 ACKNOWLEDGEMENTS

The work reported here was performed with the support of GEC Marconi Oil and Gas and the Offshore Supplies Office under grant no. 954/7177. The author would also like to express his thanks to his colleagues at MEC Peter Knight and Paul McCarthy for their important contribution to the study.

## 7 REFERENCES

[1] Cutler, C. R. and Ramaker, B. L., "Dynamic Matrix Control- A Computer Control Algorithm", AIChE National Meeting, 1979.

[2] Treiber, S. S. and Errington, S. J., "Multivariable Predictive Control of a Polymer Reactor", Proc. of CMA Conf., San Antonio, Texas, April 1992.

[3] Cutler, C. R. and Hawkins, R. B., "Application of a large predictive controller to a hydrocracker second stage reactor", American Control Conf., Atlanta, 1988.

[4] Gattu, G. and Zafirou, E., "Nonlinear Quadratic Dynamic Matrix Control with State Estimation", Ind. Eng. Chem. Res., Vol. 31, pp1096-1104, 1992.

[5] Patton, R. J., Frank, P. and Clark, R., "Fault diagnosis in Dynamic Systems: Theory and Application", Prentice-Hall, 1989.

[6] O' Mahony, S, "The modelling of subsea fluid power systems", Modelling and Control, 4th Bath International Fluid Power Workshop, Bath, 1991.

[7] Deibert, R., "Model based fault detection of valves in flow control loops", Preprints of SAFEPROCESS '94, Helsinki, pp445-450, 1994.

# athodic protection of subsea installations – a hallenge to the designer

TILLER
Norske Veritas

## 1 SYNOPSIS

Subsea production/injection systems are often located at large depths where diver/ROV intervention is expensive, so that use of an effective and maintenance free corrosion protection system is essential.

Cathodic protection has been successfully used for corrosion protection of subsea production/injection units in the North Sea, both in the British and Norwegian Sectors. In order to save weight and improve current distribution, cathodic protection has been used in combination with high quality thin film coatings, typically epoxy based. Such corrosion protection systems require no maintenance throughout the design life of the structure.

This paper discusses the challenges related to cathodic protection design for subsea production/injection systems, and give some practical solutions based on experience from several subsea installation projects.

# 2 PRINCIPALS OF CATHODIC PROTECTION

Carbon steel, like most metals, can experience both immune and active conditions in seawater, depending on the electrochemical potential of the steel surface. Normally, carbon steel corrodes in seawater (active conditions), by oxidising the iron atom into a ferrous ion and electrons:

$$Fe \rightarrow Fe^{2+} + 2e$$

This reaction that produces electrons is called the anodic reaction. On freely corroding steel this reaction must be balanced by an equivalent and opposite reaction that consumes the electrons, termed the cathodic reaction. In seawater this is usually effected by oxygen dissolved in the water:

$$H_2O + \frac{1}{2}O_2 \rightarrow 2\ OH^-$$

There will be both anodic and cathodic areas on the same steel surface.

Electrons (i.e. current) can be fed into the steel. The cathodic reaction then removes these 'extra' electrons and the anode reaction slows down, even stops or reverses. The reaction thermodynamics imply that the steel surface potential must be lowered (i.e. cathodically polarised) as the anode reaction is suppressed. The current can be supplied by impressed current generated from an external source via inert anodes (impressed current cathodic protection), or by electrically connecting a less noble metal to the steel structure, i.e. introducing an anode to the cell (cathodic protection by sacrificial anodes). In the North Sea cathodic protection by sacrificial anodes are most commonly used, since this system represents a maintenance free protection system, run by Mother Nature in contrast to impressed current systems that need closely monitoring, are not maintenance free, and need a lot more cabling.

# 3 KEYWORDS FOR A SUCCESSFUL CATHODIC PROTECTION DESIGN

Keywords for successful cathodic protection by sacrificial anodes are:

- Adequate total anode net mass
- Adequate total anode current output
- Adequate electrical continuity between anode and the steel items to be cathodically protected, both through the steel and the electrolyte.

## 3.1 Anode net mass

The total designed anode net mass shall at least fulfil the anode net mass requirements based on the current demand for maintenance of the protecting potential (mean current demand). The fulfilment of this requirement depends on the volume of anode alloy.

## 3.2 Anode current output

In addition to the anode mass, the anodes must be able to provide enough current for the initial polarisation of the steel structure (initial current demand), and polarisation of the steel in the final conditions when the coating breakdown is at its maximum and the anode surface is at its minimum size (final current demand). Fulfilment of the initial and final current demands depends on the exposed surface area of the anode.

## 3.3 Electrical continuity

Finally, the anode has to be electrically connected to the steel surfaces. The connection between the anode and the steel surface should preferably be a welded connection between the steel and the anode insert. If the anodes cannot be located directly on the steel surface to be protected, a welded connection between the metal surface where the anodes are located and the surface to be protected is preferred. Another acceptable connection is metal to metal flanges with metal gaskets. Bolted connections give acceptable continuity across the threads only, so they should be avoided unless the bolt can be fastened into the steel. Electrical continuity through other bolted connections should be ensured through cables that are either welded or brazed to each of the surfaces, or through serrated washers that 'bite' into the surface.

The absolute maximum allowed resistance over a connection between steel areas where the anode is located and other steel areas to be protected, is defined in the NORSOK standard as 0.10 Ohm. This requirement reflects the common opinion today. The maximum allowed resistance over a connection can be calculated by Ohm's law:

$$R_{max} = \frac{\Delta V}{I_{max}}$$

where $\Delta V$ is the maximum acceptable voltage drop over the connection, defined as 10% of the driving voltage between the anode potential and the potential of polarised steel (V) and $I_{max}$ is the maximum current demand for the steel surface to be protected (A), i.e. the current that must be drawn through the connection.

In addition to the electrical continuity through the metal, the anodes and the steel to be protected also need to have electrical continuity through the electrolyte. Generally seawater has good electrical continuity, and anodes work over quite long distances. However, if the seawater path is restricted through holes, channels etc., then anode effectivity is reduced accordingly.

## 3.4 Cathodic protection design philosophy

The following cathodic protection design philosophy has been the aimed at in design work:

- Each module should be self supported with anodes.
- Where anode installation is impossible, electrical continuity shall be ensured between modules and dedicated anodes.
- As few anode types as possible shall be used in order to minimise the risk of interchanges during anode installation, and to simplify anode procurement and anode production.

# 4 CHALLENGES OF CATHODIC PROTECTION DESIGN FOR SUBSEA PRODUCTION/INJECTION SYSTEMS

Now that the principals and an idealistic cathodic protection design philosophy are known, the requirements of the subsea module designer have to be met. Some of these requirements represent major challenges to the cathodic protection designer.

First, subsea production/injection systems are very compact. The size of the template is optimised (or rather minimised) with respect to space required for the subsea modules. Hence, the space available for anode installation is marginal. Very often there is a requirement regarding 'overtrawlability' which makes anode installation on the 'outboard' side of the structure impossible.

The second challenge is the complicated geometry of the subsea modules. It is almost impossible to install anodes anywhere but on the framework/support of the modules.

This leads to the third challenge: The lack of electrical continuity between the steel structures where anodes can be installed, and the modules to be protected. Bolted connections and actuators/connectors are very common. Actuators provide a good mechanical connection, but do not necessarily provide sufficient electrical connection for the cathodic protection.

The fourth challenge for anode installation is subsea tool access and access for ROV operation. 'Landing spots' for ROVs and subsea tools are forbidden areas for anode installation.

The fifth challenge is internal corrosion protection of narrow, flooded compartments such as termination heads and connectors. Available space for anode installation is minimal. Elevated temperatures caused by the presence of hot fluids have to be accounted for, as this implies increased demands on anodes. The combination of dissimilar metals, e.g. stainless and carbon steels, represent potential galvanic cells, and bi-metallic corrosion can occur.

# 5 PRACTICAL SOLUTIONS

## 5.1 'Overtrawlability', space limitations and complicated module geometry

The philosophy of using as few anode types as possible is sometimes difficult to achieve with the space limitations of a subsea template. Anode types have to be chosen carefully, and sometimes designed especially, for almost each module. Basically, smaller, flush mounted anodes are used to fit within the H-beams and I-beams of the support structures. The philosophy is here to keep the anodes within these beams, so that they do not interfere with adjacent modules.

## 5.2 ROV and subsea tool intervention

For modules that are designed for ROV and subsea tool intervention the available space for anode installation is scarce. Unless smaller, flush mounted anodes can be installed on these modules, they have to be electrically connected to other modules where anode installation is possible. Since these modules often are installed at a later stage than the modules where the anodes are located, and also being designed for retrieval, an ROV operated 'snap'-connection is preferred. However, the part of these connections that are installed first, may experience formation of calcareous deposits that will make the continuity of the later electrical connection questionable. This problem can be solved by installing an ROV retrievable protection cap for the electrical connector.

## 5.3 Ensuring electrical continuity

Measurements of electrical resistance should be performed over all bolted connections (i.e. nut and bolt connections) where electrical continuity is required. The maximum acceptable resistance should be calculated by Ohm's law. If the calculated acceptable resistance is smaller than 0.01 Ohm, it is recommended to make a connection by tack welding/brazing, or to use of a metal cable with adequate conductivity, welded or brazed to each of the modules/items. This is because measurements of resistances below 0.01 Ohm give very uncertain values.

The absolute maximum acceptable resistance between two items/modules where electrical continuity is required, is 0.10 Ohm. This value is defined by the NORSOK standard M-CR-503 (1994), and reflects the general opinion in engineering environments.

If the measured resistance is larger than the calculated maximum resistance or the absolute maximum acceptable resistance, electrical continuity must be provided through welded/brazed connections or by installing metal cables with adequate conductivity, welded or brazed to each of the modules/items.

If ROV retrievable 'snap'-connections have to be used, the part of the connection that has to be installed first should be protected against formation of calcareous deposits by an ROV retrievable protection cap.

### 5.4 Corrosion protection of internal, waterfilled compartments

In most cases, the protection of internal, waterfilled compartments is solved by installing rubber flaps to minimise the exchange of seawater within these compartments, i.e. to obtain closed compartments where the corrosion rate very soon drops down to a negligible rate because initial corrosion will consume all oxygen. In addition, biocide is added to minimise bacterial corrosion.

However, if the temperature is elevated by production fluid, making thermal convection possible, or alloyed steel is combined with carbon steel, cathodic protection is required. Very special long and thin anodes have been designed for this purpose. If these anodes cannot be used due to space limitations, it is recommended that thermally sprayed aluminium or other high quality coating is used for coating of as much of the internal surfaces as possible.

### 5.5 Other possible 'problem' areas

Finally, some attention must be given to materials and detailed design. The strength must be checked for critical components as high strength steel can suffer cracking by hydrogen released by the cathodic protection. Any potential galvanic corrosion (bi-metallic corrosion) between dissimilar metals, e.g. stainless and carbon steels, must be eliminated by ensuring adequate cathodic protection on both surfaces. Items and components with tight dimensional tolerances should be reviewed to ensure adequate functionality even with the calcareous deposit that is formed on protected surfaces.

## 6 CONCLUSIONS

### 6.1 Challenges

The challenges of designing a cathodic protection system for subsea production/injection systems can be summarised as follows:

- Compact subsea systems and difficult module geometry results in lack of available space for anode installation

- Requirements to 'overtrawlability' and ROV and subsea tool intervention implies that anodes must be 'hidden' within beams.
- Lack of electrical continuity between anodes and surfaces to be protected.
- Internal corrosion protection of flooded compartments, e.g. within termination heads or connectors, is very difficult due to space limitations

## 6.2 Practical solutions

Some practical solutions to the challenges given above are summarised below:

- Small, flush mounted anodes are widely used for subsea systems. Anodes have to be carefully chosen for each module, in order to make an optimal design
- Electrical continuity must be ensured over connections that are not welded or flanged with metal gaskets. Measurements of electrical resistance should be made over all bolted connections where electrical continuity is required. Welded or brazed metal cables with adequate conductivity is recommended for connections without adequate electrical continuity. If ROV operated 'snap'-connections are to be used for electrical connection of modules, the part of the connector that is installed first should be protected against formation of calcareous deposits by an ROV retrievable protection cap.
- Internal, flooded compartments can be protected with biocides if exchange of the trapped seawater is avoided by installation of rubber flaps. However, if the internal surfaces are heated by production fluid, or alloyed steel is combined with carbon steel, the internal surfaces should be protected by special anodes or by thermally sprayed aluminium coating or another high quality coating.

# Control Fluids and Seals

# Subsea control fluids today and for the future – the SSSV test rig

GLOBE
Castrol Offshore Limited

An increasing number of Offshore developments will involve production in the High Pressure, High Temperature (HPHT) regime. Hydraulic control still represents the most widely used means of operating subsea and surface production systems, but the conditions prevailing within the control system present new demands which have to be met by the hydraulic control media. The most significant of these are the system temperature prevailing in the subsurface safety valve within the well bore and the system pressure required to operate this valve. Test methods have been developed to establish that control fluids will provide satisfactory performance in service under HPHT conditions.

## 1 INTRODUCTION

Hydraulic fluids are used as the operating medium in virtually every Offshore production control system. As well as fulfilling their primary function of transmitting motive power, hydraulic fluids provide protection of the hydraulic control system through their lubricating and anti-corrosion characteristics. Hydraulic fluids can also interact with the control system in which they are installed in numerous ways. The degree and type of interaction that takes place within a control system can influence performance and reliability. Typical examples of such interactions are compatibility with system materials and the effect of contaminants.

There are many well-established techniques for determining the characteristics and level of performance that can be expected from a particular hydraulic fluid. These are generally described in ASTM or IP procedures and usually apply to conventional hydraulic fluids and applications. There are also test methods which are more appropriate to production control applications; some of these exist within API standards, but many have been developed by control fluid manufacturers to reflect the somewhat unique nature of the products and the environment in which they are used.

As higher temperature and pressure reservoirs are developed the performance limits of currently used control fluids are being exceeded, with potential detrimental effects on the reliability and safety of production control systems. It is therefore essential to introduce new hydraulic control fluids if HPHT wells are to be developed safely and reliably. However the current test methods available to evaluate fluid performance fail to accurately reflect all of the effects that may be introduced under HPHT operating conditions. It is as a direct result of this situation that Castrol undertook to develop a test programme that would relate as closely as possible to the in-service parameters, particularly those parts of the control system such as the subsurface safety valve (SSSV) and its control line exposed directly to the production flow within the well bore.

At the core of this programme is a test rig which incorporates typical components and materials that might be encountered in a SSSV configured for installation in an HPHT well. This enables the effects of HPHT to be studied in detail. The major benefits over conventional bench tests are that the test procedure incorporates both static and dynamic phases, thus closely emulating in-service operating conditions, and that the interaction between the fluid and the equipment can be readily evaluated. This paper describes the rationale behind the test programme and the operational aspects of the SSSV test rig. A number of issues are raised by the programme, and these are discussed. Possible future enhancements to the test programme are also suggested.

## 2 HIGH PRESSURE HIGH TEMPERATURE CONDITIONS

### 2.1 Definition Of HPHT

There have been numerous attempts to define the parameters of HPHT. A commonly quoted definition is where the wellhead shut-in pressure exceeds 10 000 lbf/in$^2$ and the flowing wellhead temperature exceeds 150°C. These conditions are certainly beyond the capability of most of the control fluids that have been in use Offshore in the past, and basic laboratory testing can demonstrate that such fluids exhibit rapid and major deterioration when exposed to these conditions.

While this definition may cover current HPHT applications, it is clear that there are numerous fields due for development which will exceed these conditions by a considerable margin, and any fluid development programme, and hence the required test parameters, should reflect the maximum conditions that might be realistically be encountered over the next few years. Castrol have therefore selected as a definition of HPHT applications where the wellhead shut-in pressure is up to 15 000 lbf/in$^2$ and the flowing wellhead temperature is up to 205°C.

Any test parameters applicable to the SSSV control system would need to reflect the above conditions, but would have to take account of the facts that the SSSV operating pressure would be significantly above the shut-in pressure and also that the temperature gradient in the well bore would expose the SSSV to a temperature somewhat higher than that encountered at the wellhead.

## 2.2 Current Experience

It has been stated on numerous occasions that there is considerable experience of operating HPHT wells and that there are a number of hydraulic fluids available which have operated satisfactorily under quite severe conditions. Further investigation suggests that if not all then the vast majority of these instances have involved either land wells or surface-completed offshore wells in relatively shallow water. Data showing the number of workovers that might be expected during production has not thus far been identified.

There is little doubt that there are hydraulic fluids available which can operate under HPHT conditions. The requirements of a control system hydraulic fluid in Offshore, and in particular Subsea, applications are much wider than simply the ability to resist degradation while exposed to high temperature and pressure. These fluids must have low viscosity to give fast system response; they should resist contamination with seawater and completion fluids; they must not be harmful if released into the marine environment; they must maximise the reliability of the production control system as a whole through their lubricating and anti-corrosion properties. It is the achievement of all of these characteristics in addition to the ability of the fluid to function under HPHT conditions that is the challenge.

There are already a number of North Sea developments just within the HPHT regime, and these are generally utilising synthetic hydrocarbon based fluids specifically developed for the purpose. These are thus far performing satisfactorily, but it should be realised that they are operating well within the design envelope of these control fluids and hence little can be inferred as to the effects of more extreme conditions.

The likely effects of exceeding the capabilities of the hydraulic control fluid have been observed where mature developments utilising conventional mineral oil hydraulic control fluid technology have experienced gradual increases in flowing well temperature over a number of years, eventually resulting in reliability problems down hole requiring workover. Conversion to synthetic hydrocarbon based fluids has eliminated these difficulties. An incidental result of studying one of these instances was a realisation that conventional mineral hydraulic oils have some characteristics which are actually unsuitable for Offshore, and particularly Subsea, operation of production control systems even at conventional pressures and temperatures. This is the subject of a separate study.

There are numerous Joint Industry Projects under way or planned which deal with both the general aspects of development of HPHT wells and specific applications such as the operation of SSSVs under HPHT conditions.

## 2.3 Future Trends

It is probable that for the foreseeable future the target HPHT limit for hydraulic control fluids defined earlier will cover the vast majority of developments. The need to develop wells exceeding these conditions is expected to be very limited, and the hydraulic control fluid is only one of a number of areas of technology that would need to be addressed before such a well could be successfully completed. In the event that a fluid were needed for this type of development then there are a number of options, but all of these would require a significant modification of the hardware involved if they were to be practical.

# 3 HYDRAULIC CONTROL FLUIDS IN HPHT ENVIRONMENTS

## 3.1 Effects Of Exposure To High Temperature

It is well understood that subjecting any lubricant to excessive temperatures over extended periods will adversely affect performance. Research suggests that a fluid can degrade in one or more ways if it is not specifically formulated with the objective of thermal stability as a main consideration. Instability at high temperature is observed when chemical changes occur within a fluid. The way in which these changes manifest themselves are as follows:

- Solubility of the additives may be reduced such that they drop out as solid or waxy deposits; this has obvious consequences with regard to the reliability of operation of critical equipment such as the sub surface safety valve.

- The performance of the additives can be reduced such that they no longer perform their functions satisfactorily; this reduces the operational life of the system components in contact with the degraded fluid.

- Acidic by-products can be formed; this is observed in most lubricants operating at elevated temperatures for significant periods, and has the main effect of increasing corrosion rates on materials with which the fluid is in contact.

In general the rate of degradation is related to the temperature, and this is an exponential relationship. Ideal relationships are however rare as most hydraulic control fluids are complex blends of chemicals which may not react in an entirely predictable way based on the expected behaviour of individual components.

High system temperatures also have an effect on the physical characteristics of any hydraulic control fluid, and while these can be minimised by the selection of a suitable product these effects must be considered during the design of the system. These effects are as follows:

- Fluid viscosity is reduced with increasing temperature. This in turn results in lower lubricant film thickness with potential for increased wear rate on sliding contact components. Changes in viscosity can modify the response characteristics of a control system, although this would be largely restricted to the SSSV and its control line within the well bore. Hydraulic control fluid leakage rates can also increase.

- Higher temperature can increase reactivity between fluids and component materials, in particular elastomers and plastics used as sealing elements within the control system. While this may be considered in the design of components within the SSSV itself it should also be noted that other components located outside the well bore but close-coupled to the flowing well stream (typically tree valve actuators) may be subjected to temperatures outside those normally encountered in conventional service, and this may require the selection of alternative sealing materials.

- Water based control fluids have a boiling point of approximately 100°C at atmospheric pressure. It is the increased pressure within the control system that allows these fluids to be used in higher temperature applications but consideration must be given to what happens if the SSSV is closed and system pressure removed. Under these circumstances only the hydrostatic head in the system above the wellhead will prevent boiling. If the head is insufficient then provision must be made in the control system design to retain the required minimum pressure on closure of the SSSV.

## 3.2 Operation In High Pressure Systems

Pressure in itself does not have a significant degrading effect on hydraulic control fluids. It is the indirect effects that need to be identified and considered, and the combined influence of pressure and temperature together. Early work suggested that lower pressures combined with high temperature would not have a major influence (*1*), but it is now believed that degradation could be induced if secondary effects prevail. The most significant aspects of high pressure in a control system with regard to the hydraulic control fluid are as follows.

- Increasing pressure within a control system will generally raise the contact loads on sliding contact components. This will decrease lubricating film thickness and raise the potential for component wear.

- Viscosity increases with increasing pressure (*2*). The effect in high temperature areas is likely to be insignificant, but may have to be considered in the control umbilical if this is of small bore and/or particularly long.

- Gas solubility increases with increasing temperature. Produced gas can migrate through fluid films across seal surfaces, and the amount of gas dissolved within the control system can be considerable. The effects of contaminating gases on the fluid needs to be evaluated. Also, if the control pressure is removed to shut the SSSV this dissolved gas will come out of solution, possibly modifying the valve characteristics. These effects will be magnified if the valve allows leakage from the well bore into the hydraulic control system.

- Fluids compress to some extent under pressure, and this can influence system characteristics (*3*). Water based fluids have lowest compressibility, with hydrocarbons being somewhat more compressible. Silicone fluids have been suggested as subsea control fluids but have a much higher compressibility and would thus be unsuitable in longer control lines.

## 4 BENCH TEST METHODS FOR PERFORMANCE EVALUATION

### 4.1 Lubricating And Anti-Wear Ability

As the conditions which prevail under different regimes of lubrication vary widely, it is impossible to use a single test method to evaluate performance. There are at least 9 identified basic test methods and a similar number of different lubrication conditions (*4*). Castrol employ several test methods which have been found to reflect in-service conditions, but many of the other methods available have no demonstrable correlation. The Shell 4-ball Lubricant Tester simulates the conditions encountered with high contact loads over relatively small areas of contact, such as might be the case at the swash-plate of an axial piston pump. The Falex Lubricant Tester is more relevant to lower contact loads, for instance the sliding element of a shear-seal type control valve. The SRV lubricant tester is a sophisticated machine which can be used to simulate a wide range of operating conditions and incorporates reciprocating motion. The main characteristics of each test method are as follows:

- The Shell 4-ball Lubricant Tester evaluates the anti-welding characteristics of highly loaded contacts immersed in the lubricant under evaluation. The test elements, as the name suggests, are ball bearings, and these are rotated against one another. The lubricant performance is determined by evaluating the load that can be achieved before welding of the test pieces occurs. Alternatively, given loads are applied for specified times, and the wear evaluated by measurement of the wear scar dimension. This test gives good correlation when applied to hydrocarbon based fluids.

- The Falex Lubricant Tester evaluates lubrication and wear characteristics on test pieces consisting of a cylindrical pin rotating between a pair of V-blocks, the test assembly being immersed in a bath of the lubricant under test. The test procedure applied is usually that proposed for the API 17F standard covering subsea fluids, and is in 2 parts; Part A involves the incremental application of load from 0 to 500 pounds, while measuring the resulting torque at each increment, and Part B involves the application of a constant load of 500 pounds for a test period of 30 minutes, recording the total wear during this period. Linear wear during the test is indicated by the number of loading wheel ratchet teeth required to maintain the 500 pound load, while total final wear is evaluated by measuring the weight loss of the test pieces. This test gives good correlation when applied to water based fluids.

- The Optimol SRV Lubricant Tester is used to evaluate the extreme pressure performance of a fluid by measuring the coefficient of friction with increasing load. The test regime differs from that employed in the Falex test in that the test piece configuration is a ball sliding on a plate, and the motion is reciprocating rather than rotating. The short sliding distance and rapid cycle time thus establishes that the fluid provides good lubricating ability even when conditions prevent the formation of an hydrodynamic film. This test provides good correlation with practical applications such as shear-seal directional control valves and all-metal sealed subsurface safety valve actuators.

Good performance in more than one lubrication regime is generally necessary, and good performance using one test method does not necessarily infer good performance in other regimes. All of the test methods described above can be modified to study specific situations, but none of them can effectively introduce the important parameters of high lubricant pressure or temperature.

Pump testing is also widely used to evaluate the performance of hydraulic fluids, but as yet there are no widely recognised standards for such testing that cover the parameters experienced in production control systems. Evaluations tend to be run on a case by case basis.

## 4.2 Anti-Corrosion Performance

One of the most basic (and probably one of the most severe) procedures is the IP135 corrosion test on carbon steel. This involves contaminating the control fluid with water and immersing a steel pin in the contaminated fluid, which is heated and agitated. To pass this test there must be no detectable corrosion with both 10% distilled water contamination and 10% seawater contamination. This test procedure can be modified to evaluate effects on differing materials.

Because of the wide variety of metals encountered in modern systems, extensive corrosion testing is carried out to determine the effects of galvanic corrosion. Metals addressed include carbon steel, Monel 400, Incoloy 800, aluminium bronze, copper nickel, stainless steel and cast iron. Multi-metal systems are evaluated on the BK Technology Digi-Galv Corrosion Analyser. This device allows multi-metal systems to be evaluated in a fluid under controlled conditions, with data on corrosion potential being gathered in real time. Castrol utilise a special test head incorporating materials employed in the manufacture of production control systems .A corrosion potential of below 1 microAmp per sq. cm. is considered to be passive. Materials exhibiting a negative corrosion potential are actively protected. Galvanic corrosion is generally associated with water based fluids but must also be considered if water contamination of hydrocarbon based fluids occurs.

## 4.3 Thermal Stability

Thermal stability is studied by holding the fluid at elevated temperatures for extended periods. There are numerous recognised test methods which evaluate stability (tests for oxidative and hydrolytic stability are examples) but few of these involve parameters which relate very closely to the in-service conditions. They may be used as indicators of relative performance, but the data obtained should be treated with care, particularly as they are generally designed to give a result in a short period of time, typically a few days.

Realistic testing involves much longer time scales, with experience suggesting that a minimum of eight weeks is necessary to show up the effects of exposure to high temperatures. The methods used to carry out such tests vary widely, but the more closely the in-service condition is simulated the better the correlation obtained. Conversely of course the cost and difficulty of carrying out a test increases sharply with more sophistication and a balance needs to be struck. The best approach is to use less sophisticated methods to screen candidate products and to reject any which clearly do not meet the criteria, and then to progress to more advanced testing. Two screening methods appropriate for screening before moving on to a full test on the SSSV test rig are used:

- Atmospheric pressure testing is carried out in stainless steel vessels. These are filled with the fluid under test and placed in controlled temperature ovens. The vessels are designed to be easily accessed for examination and cleaning. In the procedure used by Castrol, no attempt is made to exclude air. The fluid condition is evaluated after test periods of up to one year.
- Hyperbaric testing is carried out in stainless steel pressure vessels. These are pressurised to a level at or slightly above the maximum operating pressure and heated to the test temperature.

Atmospheric testing has also been carried out with the exclusion of air by the introduction of nitrogen into the test vessel. There is some evidence that this may create an environment more benign than that encountered in service and it is felt that the inclusion of air is more likely to show up any thermal instability in a fluid and hence avoid a failure during evaluation on the full SSSV test rig..

## 4.4 Materials Compatibility

Fortunately there are many recognised and appropriate test methods for evaluating the compatibility of materials - metallic and non-metallic - with fluids. Most of these are relatively easy to carry out. In most cases it is preferable to carry out compatibility testing for Offshore applications for longer periods than might be normal in general industry to establish that there are no long-term effects that might not become apparent over a shorter period.

Test methods on elastomers and plastics involve the testing of standard test pieces of the material. Experience however suggests that these can give results significantly different from those obtained on actual components. It is therefore advisable to test on the components themselves. Most of the ASTM test methods for compatibility testing allow for this, and also can incorporate the use of longer than standard test periods.

## 4.5 Limitations Of Bench Test Methods

While most selected bench test methods demonstrate performance of a fluid under various conditions, none of them incorporate all of the conditions encountered in service. Additionally it is not possible to accurately determine what the effect of interaction of the various conditions and materials might be in combination. It is for this reason that Castrol decided to develop a method which would incorporate as many of these conditions and materials as possible into one test rig which would allow the testing of fluids in such a way as to give a high degree of confidence in the product, and also provide valuable data for the ongoing control fluid development programme. The key features were to be the incorporation of both temperature and pressure, the use of materials found in the SSSV itself and a method of evaluating the lubricating performance of the fluid under these conditions.

## 5 SIMULATING IN-SERVICE CONDITIONS - SSSV TEST RIG

### 5.1 General Description

The test rig is a self contained unit incorporating a SSSV test element together with all power and control ancillaries required to simulate in-service conditions (*Fig. 1*). The test element is electro-hydraulically controlled. High-pressure nitrogen is used to simulate the well-bore pressure. An electrically heated jacket provides the high temperature environment. A control and data acquisition system automatically implements the pre-defined test profile and records relevant measurements made during the test sequence. The test rig in its current configuration is rated for a maximum safe working pressure of 17 000 lbf/in2 and a maximum operating temperature of 200°C. These limits can be increased with appropriate modifications.

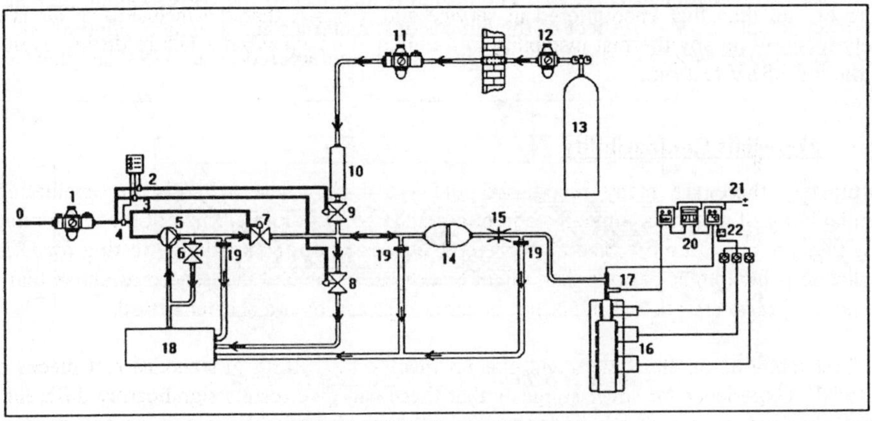


*Fig. 1 SSSV General Layout*

| | | | |
|---|---|---|---|
| 0 | Air supply (110 lbf/in$^2$) | 12 | Nitrogen regulator |
| 1 | Regulator with filter | 13 | Nitrogen cylinder |
| 2 | Valve solenoid A | 14 | Fluid transfer chamber |
| 3 | Valve solenoid B | 15 | Flow restriction |
| 4 | Valve solenoid C | 16 | Temperature controller and heaters |
| 5 | Air drive pump | 17 | Pressure controller and transducer |
| 6 | Manual shut off valve | 18 | Oil reservoir |
| 7 | Normally closed pneumatic valve | 19 | Pressure bursting disks |
| 8 | Normally open pneumatic valve | 20 | Chart recorder |
| 9 | Normally open pneumatic valve | 21 | 240 V supply |
| 10 | Pressure Intensifier | 22 | Transformer 240V to 50V |
| 11 | Nitrogen regulator with filter | | |

*Key to Fig. 1*

The test rig is designed to be operated on both synthetic hydrocarbon and water based control fluids. The different viscosity characteristics of fluid types are compensated for by provision of a metering valve in the control line emulation component assembly which eliminates the need for long control line runs and enables the valve response to be similar to that found in service.

The SSSV test rig is located at the Castrol International Technology Centre at Pangbourne, Berkshire.

## 5.2 System Components

The major components and their functions are described in the following paragraphs:

- The ***SSSV Test Element*** is a test block incorporating a hydraulic actuator and return spring arrangement (*Fig. 2*). The design is identical to the corresponding actuator in an actual SSSV in respect of the dimensions, materials and sealing configuration. This component would normally be designed and manufactured by an SSSV supplier.

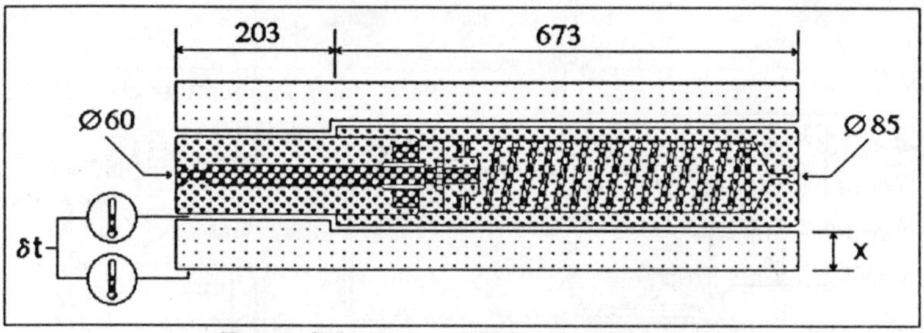


*Fig. 2 Schematic of SSSV Test Element*

- The ***Hydraulic Power Unit*** uses a pneumatic pump and nitrogen pressure intensifier system to provide hydraulic power to the SSSV Test Element. High pressure nitrogen is also used to simulate the well bore pressure at the base of the SSSV test element.

- The ***Control Line*** is emulated by two components This eliminates the difficulties of providing an extremely long vertical control line. The fluid volume is prevented from mixing with the remaining control fluid in the system by a fluid transfer chamber incorporating a floating piston. This ensures that as is the case in service the same fluid is used repeatedly to operate the SSSV. The flow characteristics of the control line are simulated by passing fluid through an accurately calibrated metering valve.

- The ***Control And Data Acquisition System*** regulates the test rig temperature and pressure to those parameters set for the test run and monitors and controls the test sequence (*Fig. 3*).

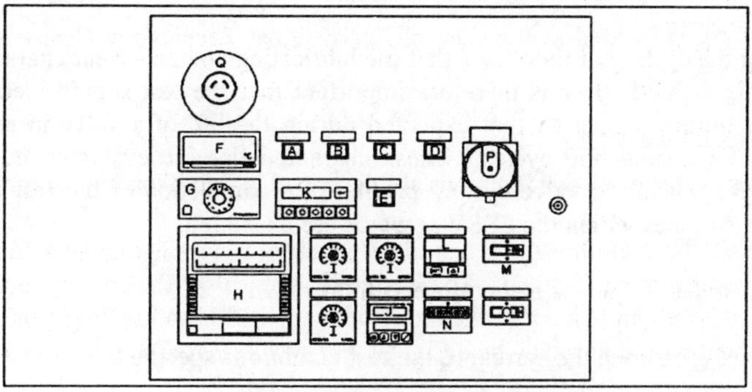


*Fig. 3 Control Panel*

| | | | |
|---|---|---|---|
| A | Latch switch - Charge system | J | Autotuning temperature controller |
| B | Latch switch - Discharge system | K | Pressure controller |
| C | Latch switch - Stagnant system | L | Actuation cycle counter |
| D | Latch switch - Start actuation cycle | M | Elapsed time meter |
| E | Latch switch - Start/Stop Heaters | N | Actuation cycle limit |
| F | Thermocouple temperature display | O | Actuation cycle timer |
| G | Thermocouple switch selector | P | Emergency stop button |
| H | 100mm strip chart recorder | Q | Safety control isolator switch |
| I | Temperature limit control | | |

*Key to Fig. 3*

Separate control functions are provided for charging the system, holding the system in a static condition, cycling the SSSV Test Element actuator for a "block" of cycles and discharging the system. The actuator cycling profile can be pre-set to control independently the valve open and closed times and the number of cycles per "block". Control is by solenoid operated pneumatic valves and data acquisition by thermocouples and pressure transducers. A high speed chart recorder monitors the pressure and temperature profiles during valve actuator cycles. Safety is provided by interlocks, fail-safe system discharge in the event of power failure and bursting discs.

## 5.3 Test Procedure

In order to closely simulate the in-service conditions encountered in the operation of an SSSV in an HPHT well it is necessary to include both the static and dynamic parts of the SSSV function. To ensure that any effects of the test conditions were able to be observed it was decided to adopt a 12 week total test period.

During the static period the valve is held open under pressure, and it and the fluid are subjected to the flowing well temperature. This is the most severe condition to which the fluid will be subjected. SSSVs are generally operated periodically in service, and it was decided that a four week static period would be appropriate.

It is during the cycling of the SSSV that the lubricating and anti-wear characteristics of the fluid will be highlighted. It was therefore important that the test should incorporate a total number of operations similar to that expected during the life of a valve in service. This is anticipated to be around 600 cycles. These characteristics are evaluated during testing by monitoring leakage and the valve cycling profile. On completion of the test the sealing and sliding contact surfaces within the SSSV actuator are examined.

The complete test sequence is therefore as follows:

- Heat and pressurise the system to the test conditions specified.
- Operate the actuator for a block of 150 cycles.
- Leave the system static at the specified temperature and pressure for a four week period.
- Repeat the cyclic and static stages twice more.
- Operate the actuator for a final block of 150 cycles.
- Terminate the test, strip down and examine the test element and samples of the control fluid.
- Report.

It has been decided that the first of the test sequence will be carried out at a pressure of 10 000 lbf/in$^2$ and a test temperature of 170$^{\circ}$C. The control fluid selected is Castrol Brayco Micronic 864 HT/200, a synthetic hydrocarbon based product specifically designed for HPHT applications which has performed well in screening tests.

## 6 FUTURE TEST PROGRAMME

The next test sequence will probably be an evaluation of a water based control fluid - Castrol Transaqua HT - at 150$^{\circ}$C and 10 000 lbf/in$^2$. The test profile will be identical to that already described. These first two tests will cover the conditions expected in most applications in the next two to three years.

The next stage will be to progress towards the higher end of the HPHT range described earlier. This testing will concentrate initially on synthetic hydrocarbon based fluids, and screening tests carried out thus far indicate that it should be possible to achieve at least conditions of 17 500 lbf/in$^2$ and 200$^{\circ}$C with current Castrol technology.

Once it has been successfully demonstrated that control fluids can be made to operate under HPHT conditions it is planned to introduce further factors which could have an effect on SSSV performance and reliability. These are mainly related to external contaminants that can enter the hydraulic control system. Typical contaminants that have been identified in existing control systems are:

- Produced fluids. It is well documented that hydrocarbon liquid and gas can and do enter the SSSV hydraulic control circuit, often migrating through to the surface control panel. It is necessary to establish the effects that this might have on the control fluid when combined with HPHT conditions.

- Completion fluids. Brines in particular are found in SSSV control lines. These may or may not be compatible with the control fluid, and the effects of these contaminants need to be studied. Reduction in solubility could well produce solid precipitates from the brines which could block the SSSV control line if present in sufficient quantities.

- Particulate matter. No control system is free of solid contamination, and however clean the system is at installation there will inevitably be a build up of material which could cause wear of the sliding components in the SSSV.

The design of the SSSV control circuit in a production control system is unique in that - with a few exceptions - it is a vertical dead-ended line with no means of flushing it out without costly workover of the well. It is therefore considered vital that the effect of contamination is included in any study of this type.

## 7 DISCUSSION AND SUMMARY

The SSSV test rig is in operation and initial test results will soon be available. It is felt that it will provide a useful tool in the development of control fluids for HPHT operations, as well as providing data which will be of interest to manufacturers of SSSVs.

During the development of this test rig there was a considerable study of existing test procedures and practices, and a number of interesting points were highlighted which are worthy of mention. It is hoped that readers of this paper will be encouraged to comment and that discussion on this important aspect of production control is stimulated:

- The methods currently used for evaluating lubricant performance are not adequate for proving control fluids in SSSVs under HPHT conditions. The test rig developed by Castrol represents one way of gaining more representative performance data, but as yet cannot provide all of the necessary answers.

- There is currently very little data on SSSV performance under controlled conditions. Most testing is carried out over relatively short periods, and is normally only cyclic in nature. This does not take account of the long static periods that are usual for the application.

- Although most SSSVs are installed in clean conditions - NAS 1638 Class 6 is the usual standard applied - there is little information on the effect of high levels of contamination on the reliability of these valves.

- There is a need for a forum whereby SSSV and control fluid manufacturers can discuss their respective requirements such that the valves and fluid can be treated as an interactive system rather than discrete elements.

To summarise, it was felt that current bench test methods for the evaluation of hydraulic control fluids did not accurately reflect the conditions experienced in HPHT applications. The SSSV test rig was designed to simulate these conditions more closely, and to enable evaluation of fluid and valve together. The rig is now in operation and is performing well. Enhancements are planned to extend the test envelope and to introduce additional parameters to move closer to the actual conditions that might be encountered down hole. It is hoped that in the future it will be possible to submit a further paper which will present the data obtained from the test program.

## REFERENCES

(1) GLOBE, A.S. Hydraulic Operation of Subsurface Safety Valves in High Temperature Wells, Subsea Control and Data Acquisition Conference, London, UK, 29-30 April 1992.

(2) NEALE, M.J. Tribology Handbook, (Newnes-Butterworths).

(3) GOODWIN, A.B. Power Hydraulics (Cleaver-Hume Press)

(4) TOURRET, R and WRIGHT, E.P. Performance Testing of Hydraulic Fluids (Institute of Petroleum).

# ydraulic fluids and their future in a more nvironmentally conscious industry

IAYLOR and S McMANUS
ırtson Bentley

The latest major developments in subsea hydraulic fluids have been a result of the increased temperatures and high pressures now involved in the more recent subsea wells. Some initial work has been completed and there are now fluids on the market which venture towards solving the problems of controlling these hotter wells.

Further work is being carried out in this field to raise the temperature limitations even higher, however the major change that will influence formulations and the use of hydraulic fluids up to the year 2000 will be the demanding environmental legislation and the desire for the offshore industry to show its commitment to preserving the sea and marine ecosystems.

This paper will endeavour to describe the qualities needed in an "environmentally sound" fluid and the testing that needs to be done to prove the fluid properties to the operators and the governments who manage the waters where the products are discharged.
Many thousands of pounds and man hours have already been spent in producing new fluids. These fluids are about 100 times less toxic than already "environmentally acceptable" hydraulic fluids.

The paper will give a description of how the old fluids and the new fluids can affect the environment and give an indication of how they perform. There will also be outline of the ideas of how the mechanical systems could be altered to accommodate the new fluid technology and operate under the high temperatures and high pressures required in recent and future offshore installations.

## 1 INTRODUCTION

The last couple of years has seen increased awareness in environmental issues with pressure groups bringing to the attention of the media and thus to "Joe Public" the concerns arising from industrial activity and the effect such activity has on the planet and it's environment. "Greenhouse Effect", "Depleting Fish Stocks", "Acid Rain", "Destruction of Rainforest".

The Oil Industry, in particular, has often been attacked for its past record of spillage and discharges and the damage that these have caused to the environment. For example, it is necessary only to look at the reactions following the media coverage of the Brent Spar to show that it is an issue that raises emotions high.

The Offshore Industry in the North Sea has seen a notable change in the legislation applicable to the discharge of chemicals into the marine environment. The legislation limits the quantities of fluid that can be discharged into the sea depending upon the chemical content of these fluids and the manner in which they are used.

It is as a consequence of such legislation that subsea control and drilling mud operators are having to look at chemicals disposed of during routine operation and evaluate the possible effects that they cause to the environment.

The advance of technology for high temperature and pressure wells has also been a major contributor in current developments offshore with producers now looking to extract crude from the higher temperature wells and looking for the ways and means in which to do this.

The design of the control systems and the hydraulic fluids used in these systems is thus being considered in great detail by the relevant companies.

The producers are looking for systems that are capable of functioning at the higher well temperatures and discharging lower volumes of fluid or are having to consider the use of closed loop systems whereby the fluid is returned to the platform from the wellhead, a more costly option requiring extra equipment to operate effectively.

Alternatively, they are looking at recent developments in fluid technology for fluids that are not only stable at the higher temperatures but are also shown to be more chemically acceptable to the marine environment and are classified by the regulatory bodies as such.

Fluids are continually being developed to meet the changing demand from the fluid users. As a result of the increased awareness and corresponding requests from the industry for "greener" products, new fluids, and indeed those already existing, are having to undergo stringent testing in order to evaluate their probable behaviour when discharged offshore.

These fluids not only have to be able to pass the limits set for the environmental testing but which looks at toxicity values using a variety of marine organisms, biodegradability and bioaccumulation potentials but also require the properties necessary for their purpose in operation, which in the case of the subsea control fluids means they need to have a low viscosity for fast system response, they require good lubricating properties and anti-wear for running through pumps and valves, they also should possess corrosion resistance properties to increase the life of the plant equipment. Not only all these but also the thermal stability of the fluid is important, especially considering the development and production in the high temperature wells of more recent history.

## 2 CONTROL SYSTEMS

Subsea control systems are generally put into two groups, closed loop and open systems. Both systems are constructed from the same materials, metals, seals and hoses the difference is in what happens to the control fluid used to operate the valves.

The terms are fairly self-explanatory, closed loop systems contain return lines with all the fluid being returned from the wellhead up to the platform. The extra equipment involved makes the closed loop system by far the most expensive option of the two, especially with the control lines covering even greater distances as subsea systems become linked up to existing production facilities. This leads to a need for more powerful pumping systems to return fluid to holding tanks at the platform level.

Open loop systems discharge hydraulic control fluids directly out into the sea when the valves are operated. On the whole, open loop systems tend to run on water/glycol fluids and closed systems are used where hydrocarbons are the control medium although there are exceptions and there are some cases of hydrocarbons being discharged and water/glycols recirculating through closed loops. The equipment top side for hydrocarbon fluids is more expensive, as the oil not only requires filter equipment to remove particulates from the fluid, a necessity for all fluids, but also needs drying equipment to remove water from the oil before it is used. Water can enter the system from condensation on the side of holding tanks due to temperature changes. A dryer could cost anything between £40 000 to £50 000.

Open loop systems are thus cheaper as the quantity of equipment required for the system both at wellhead and topside is minimised and the extra fluid cost due to the discharging losses is far less than the closed loop system equipment. But as the fluid is discharged directly into the sea it is perhaps the less popular option with legislative authorities looking at the output of chemicals into the sea.

The figures (Fig 1 and Fig 2) below give a typical set up for a control valve, designs do vary between systems.

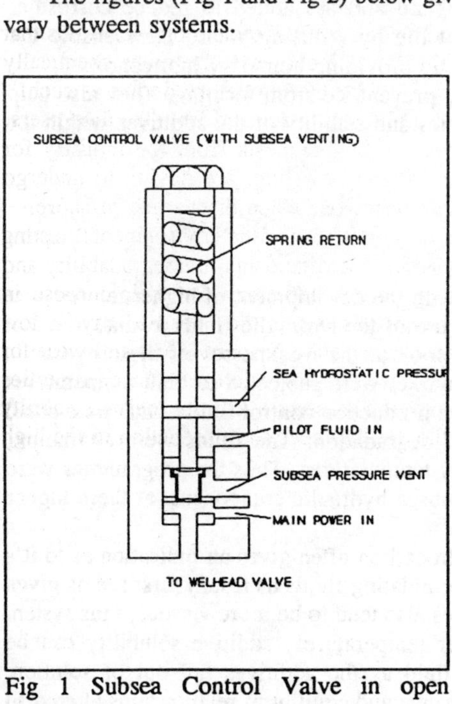


Fig 1 Subsea Control Valve in open position

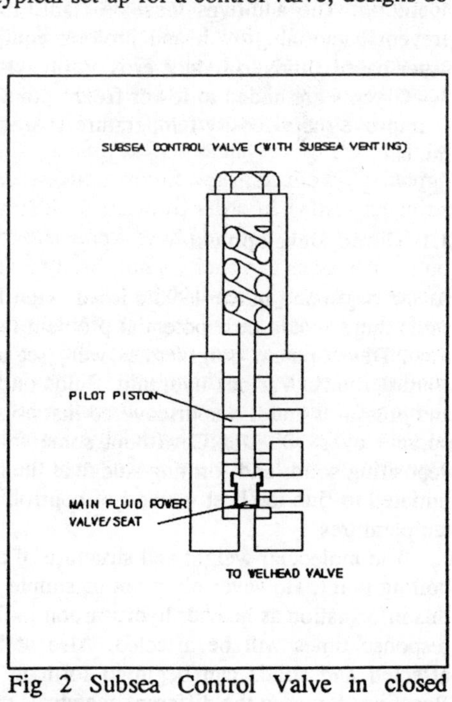


Fig 2 Subsea Control Valve in closed position

The pilot fluid works against the sea hydrostatic pressure and the spring force to open the valve allowing the main power hydraulic fluid to the wellhead valve. When the pressure from pilot fluid is released then the force in the spring and the hydrostatic pressure close the valve and the fluid from the wellhead valve is released through the subsea pressure vent either to be discharged into the sea, if it is an open loop system, or alternatively into a manifold for closed loops. The actual quantity of fluid discharged depends on the design.

## 3 CONTROL FLUIDS

It has been mentioned previously that both water/glycol mixtures and hydrocarbons are used as hydraulic medium offshore.

Hydrocarbon-based fluids are generally used in closed systems but there are some open loops operating with hydrocarbons, conservative engineers have probably specified hydrocarbon-based fluid in the past and have developed a personal preference, water/glycol mixtures have been known in the early years to have had problems with corrosion and wear, since these times though fluid formulations have improved and these problems alleviated.

On the whole water-based fluids are preferred for use on control systems offshore mainly due to safety reasons as hydrocarbons are flammable. But also the response times for water-based fluids are more rapid due to lower viscosities than the hydrocarbon fluids.

Subsea control fluids require necessary properties in order to operate effectively.
The fluids are required to be in service for the life of the well, up to 25 years in some cases. Some installations are controlled from great distances, with low viscosities thus required to reduce response times. Water has a very low viscosity thus would be ideal except that the fluid also needs to be a good lubricator and able to reduce wear on a system, water alone can be very erosive/corrosive. The temperatures that the fluid is exposed to would also be a problem, particularly topside in the North Sea, as freezing may occur.

Additives are added to the water to enhance it's properties and create fluids ideal for this operation. The additives must be stable in the fluid and are added to reduce corrosion, prevent bacterial growth and prolong equipment life by reducing wear. They enable the water-based fluids to reduce erosion on systems by providing boundary lubrication.

Glycols are added to lower freeze points and prevent ice from forming. They also help to improve the viscosity/temperature characteristics and stability of the additives within the fluids.

### 3.1 Fluid Development

At the beginning of the 1990's it was seen that with the development of higher temperature wells there would be a potential problem in the use of the hydraulic fluids available at that time. Development programmes were set up to look at the weaknesses of these hydraulic control fluids. Various hydraulic fluids on the market were subjected to high temperatures and pressures and it was discovered that no subsea production control fluid of that time could operate over 100 Deg.C without some form of degradation. The fluids were degrading, depositing solids and forming sludge at the higher temperatures. Thus the programmes were initiated to find the best method of controlling subsea hydraulic components at these higher temperatures.

The molecular weight and structure of a hydrocarbon often gives an indication as to it's boiling point. However all is not so simple in formulating fluids as it may first seems given this information as heavier hydrocarbon molecules also tend to be more viscous, thus system response times will be affected. Also at higher temperatures, additive solubility can be affected and solids can begin to form in the fluid as the additives fall out of solution. Reactions between the different chemicals can change and equilibrial relationships altered so that the properties of the fluid vary as the temperature is raised. This is true for the water/glycol fluids also.

As a result of the work carried out both synthetic hydrocarbon and water/glycol fluids have been concocted that are able to withstand higher temperatures than the fluids used offshore at present. A joint industry project was set up to simulate the high temperature and pressures involved in subsea safety valves to test the properties of these fluids.

The second issue that has had to be addressed with respect to the development of fluids is that regarding the discharge of chemicals into the sea and the resulting effect that the chemicals will have on marine life.

## 4 ENVIRONMENTAL TESTING

The discharge of fluid offshore in the North Sea is regulated with each government being responsible for the limits set in their sector. The UK waters are managed by the Ministry of Agriculture, Fisheries and Food (MAFF) with the Department of Trade and Industry (DTI) certifying the discharge limits. The Norwegian Authority responsible for evaluating the limits for discharge is SFT (Statens Forurensings Tilsyn). The Dutch Authority, The Institute of Mines, regulates their waters.

All these Authorities now follow the same set of guidelines which are known as CHARM and use the HOCNF _ Harmonised Offshore Chemicals Notification Format, these have resulted from earlier PARCOM (Paris Commission) guidelines somewhat modified and restructured.

The test work that is necessary for the evaluation of the chemicals under CHARM is outlined, fluids undergo these tests and results are submitted before discharge permits are granted. Whereas in the UK in the past, fluids have been classified into different groups with discharge limits. It looks, at the present moment, as if cases will be evaluated on a project by project basis and permits given accordingly.

### 4.1 CHARM

The CHARM Model evaluates the environmental risk for offshore chemicals by looking at hazard assessment, risk analysis and risk management procedures. The model has been developed to bring the harmonisation of differing guidelines from government regulations for the discharge of chemicals. It has been derived from a variety of principles and models used in the different North Sea countries, these include:-

"Esther" the ranking system from Scandinavia
"Prisec" hazard evaluation system from The Netherlands
OECD - workshop conclusions
"ECETOC" methods of environmental hazard assessment.

The model takes the specified data requirements,(ie. from toxicity tests, bioaccumulation and biodegradation tests and health and safety data sheets) the data is run through the model to give information which can be used as a basis for making decisions. Thus CHARM is a tool to support evaluation and does not make decisions itself.

The model looks at the comparison of estimated PEC's (Predicted Environmental Concentrations) and the estimated NEC (No Effect "threshold" concentration), using data on predicted usage, release, dilution and the fate of the chemical. The NEC uses toxicity data from agreed OECD principles. The model looks at "worst case" situations during normal usage.

In general if PEC > NEC the ecosystem is likely to be affected by the chemical tested, whereas with PEC < NEC the ecosystem will probably not be affected. The probability of effects occurring is a non- linear function of the PEC:NEC ratio.

The CHARM model uses PEC's and NEC's for three different parts of the marine system looking at pelagic (in the water), benthic (on the sea-bed), and in the food chain systems and thus analyses the separate risks in each phase.

**Risk analysis is then extended using risk management. Which looks at the measures taken to reduce risk. Risk management sets priorities to lowest hazards, prescribes environmental care dependant upon the risk (based upon BEP - best environmental practise), and identifies those measures that can reduce risk cost-effectively.**

## 4.2 Toxicity Testing

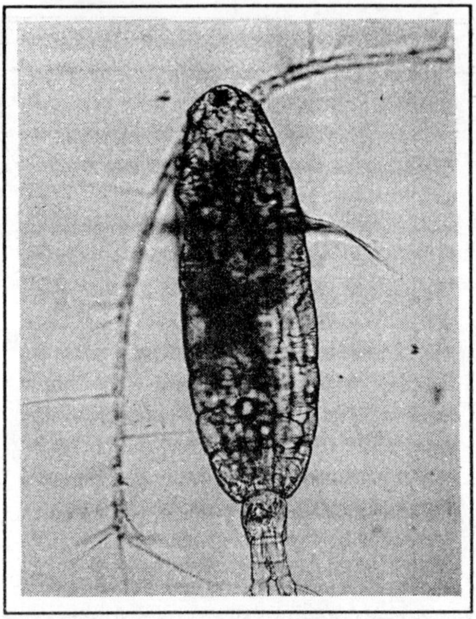

Fig 3 Marine copepod - *Acartia Tonsa*

Toxicity tests are conducted on a number of different marine organisms to assess the effect the fluid chemicals may have within the different marine phases, upon the sediment and in the water as well as in the food chain.

A marine algae such as *Skeletonema Costatum* is used in an acute toxicity test to assess the dosage of chemical required to inhibit the algal growth. This test measures the dosage required to inhibit the growth of a set percentage of a batch of the algae within a set time period following a PARCOM protocol.

Testing is also carried out on a marine copepod, for example *Acartia Tonsa* (which is pictured Fig 3), this small marine animal like the *Skeletonema* is present in the pelagic phase. Following the protocol from PARCOM, the lethal dose of fluid is measured that kills off 50% of a batch of *Acartia* within a set time limit.

For substances that are that sink or are insoluble in the sea water and therefore may collect on the sea bed, tests are conducted on an organism that lives on the sediment, the amphipod *Corophium Volutator* (pictured Fig 4). The tests are conducted in line with PARCOM protocols.

For those substances that are soluble in the sea water toxicity tests are conducted on fish. Toxicity tests may also be carried out on other organisms for example, oyster larvae.

The regulations require that these tests are carried out according to the guidelines and must be conducted at approved laboratories.

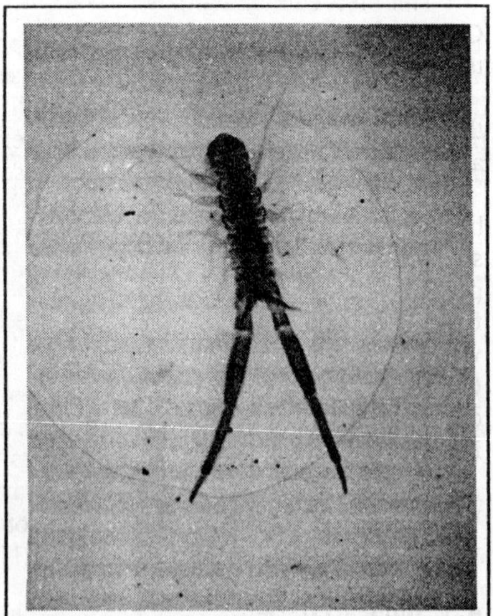

Fig 4 Sediment reworker - *Corophium Volutator*

### 4.3 Typical Toxicity Results for fluids

As the regulations have changed so have the adopted test methods. Thus some of the test results we obtained for fluids 2-3 years ago have been conducted using different test methods or different organisms therefore direct comparison of results can be difficult. However the table below (Table 1) gives an indication of the kind of results we obtain. The "older fluid" in the table has always been considered a relatively "friendly" fluid since it was developed about 10-12 years ago.

Table 1 - Typical Toxicity Results.

| TEST | Older fluid | Newer Fluid |
|---|---|---|
| *Skeletonema Costatum* or equivalent | 108 mg/l | 1069.9 mg/l |
| *Acartia Tonsa* or equivalent | 117 mg/l | >10000 mg/l |
| *Corophium Volutator* | 12037 mg/Kg | 16303 mg/Kg |

### 4.4 Bioaccumulation Potential

The legislation with respect to bioaccumulation and biodegradation now requires that all chemicals within a fluid formulation be assessed for their ability to bioaccumulate and\or biodegrade. That is are the chemicals likely to pass through biological membranes into cells of marine organisms and from then on into the food chain. Alternatively are the chemicals able to be broken down into their elemental constituents.

The bioaccumulation potential of the fluid chemicals is required in the calculations of CHARM as it helps in the determination of the probability of marine organisms absorbing these chemicals into their systems.

If the molecules within a fluid are of a relative molecular mass of greater than 600 then the molecules are assumed to be too big to pass through the pores of a biological membrane and therefore are unable to bio-accumulate.

The bioaccumulation potential is estimated from the log $P_{ow}$ (the partition coefficient between octanol and water). This is a measure of the chemical's ability to pass from one solution to the other. The idea is that the test gives an indication of whether the chemical will pass into the fatty tissues of marine organisms from the sea water. There are various methods from OECD guidelines. It can be done using HPLC (High pressure liquid chromatography) where an octanol based compound is contained upon the chromatography column packing and the water is used as the eluent phase fluid. The test looks at the retention time of the chemical in the column. If the chemical is absorbed into the octanol based compound then it is likely that the chemical will bioaccumulate.

Bearing this in mind for the fluid to pass the environmental testing the fluid formulation requires that chemical additives favour solution in water to solution in octanol.

### 4.5 Biodegradation

The ability of the chemical to biodegrade in sea water is one of the factors required of the fluid for it to pass the environmental legislation, and this is one of the more difficult criteria to meet. The ability to degrade in many cases also indicates a weakness in stability. In a search for fluids that are stable at higher temperatures, producers look for chemical structures that are unlikely to break down or react with the extra energy in the system.

The stability of a control fluid is essential especially with time. With the longer lines to the subsea valves, the time periods for fluid in a control system can be anything up to 25 years. Degradation can lead to solids forming in the system which in turn can lead to blocked valves, erosion, fluid acidity and thus corrosion, and ultimately to a safety problem if the safety valves cannot be shut.

Thus we want the fluids to degrade in the sea but this is the last thing we want to happen in the control system.

Biodegradation can be measured in a variety of ways and results have been known to differ widely even using the same test method. Most methods compare the theoretical oxygen demand, determined from the chemical structure to the actual oxygen taken up in a measured test using micro-organisms and compare with BODs (biological oxygen demand) of standard chemicals.

It is difficult to compare results of biodegradation on different fluids as laboratories use different test organisms depending on where they come from (some laboratories employ micro-organisms from sewage sludge that will have seen a wide variety of chemicals in the past and thus may be somewhat immune to more toxic chemicals), some test methods shake the test flask, this can disperse the oxygen supply more efficiently and also the test chemical thus perhaps providing a greater surface area for microbes to work upon thus enabling chemicals to break down more quickly. In other tests a closed bottle is used.

For this reason the test method that has now been adopted by the regulatory authorities as the standard is the OECD 306 test method. This is a sea water test that assesses the capabilities of the chemicals to biodegrade using the populations of bacteria that occur naturally in sea water using no added inoculum. All the different components of the fluid present at greater than 1% are required to be tested. This gives the authorities a better idea of the mechanism of how the fluid may break down when discharged.

Biodegradation tests are normally run for a 28 day period, guidelines have now changed and the test period is now being extended to 90 days if chemicals have not degraded by >60% in the 28 day period. A chemical degrading too quickly could cause more problems than a slowly degrading chemical in that it may take up all the dissolved oxygen from the water as it degrades and starve any organisms in the surrounding water of their oxygen supply while doing so.

## 4.6 Effect on the Environment

There is a theory that some marine organisms actually thrive rather than the reverse in areas around discharge vents, where they can feed off the bacteria that are breaking down the fluid chemicals. If these chemicals were toxic then there may be problems if not in the short term then maybe longer term if the chemicals are liable to bioaccumulate and build up in the food chain. Algal growth can be reduced this would lead to a cut down in the food source for other marine species and thus upset the delicate balance within the ecosystem.

Some chemicals will have more of an effect than others and there are lists of chemicals that have been banned from use in recent years or are very closely monitored by the authorities.

So these fluids are tested for toxicity, biodegradation, and bioaccumulation to test their likely affect on the environment. The CHARM calculations look at the dilutions upon discharge and it is based upon the CHARM modelling that authorities make decisions upon what can or can not be discharged and in what proportions. The current guidelines have only recently come into play and therefore it is difficult to say at present how the different governing bodies are applying them.

## 5 CONCLUSIONS

There are control fluids that have been formulated and are shown to be less toxic. A great deal of money and time has been spent in testing formulations according to the guidelines. Over the next few months we will be able to evaluate more fully the market performance and thus the interest from operators.

With the risk management criteria as part of the guidelines it looks as if the more recent less toxic fluids may be favoured in the future but it may depend on how far legislation goes in backing the guidelines that have been set. Ultimately fluids may have to be contained in the more expensive closed loop control systems whether toxic or not.

With all the environmental aspects of the fluids becoming more and more an issue to consider when deciding upon a control fluid, it is also important to keep in sight the purpose for which these fluids are designed. They are used in the control of safety valves, thus they may themselves be able to prevent an environmental disaster. It is therefore important when developing new fluids not to compromise the properties that we have already described as necessary for control fluids to have;- Stability, especially in view of the higher temperature and pressure wells. Low viscosity to keep response times to a minimum, especially considering the longer distances for control lines to platforms. Good lubrication to reduce wear and prolong the life of a system. Corrosion inhibition and biocides to protect equipment and additives to reduce freezing points and prevent ice from forming.

## 6 ACKNOWLEDGEMENTS

The authors wish to thank the following people who have assisted in the preparation of this paper:- Alan Hunt of Kvaerner FSSL Ltd - for general information regarding control systems.

ERT - Environment and Resource Technology Ltd - for photographs of organisms.

# System Integration

# Experience from design and commissioning of integrated subsea and platform control systems

**CORNELIUSSEN**
ongsberg Offshore as
**HAUGAN**
koner Engineering as

## ABSTRACT

The design philosophy of Subsea Control Units (SCU) (topside electronic equipment for controlling subsea wells) have changed significantly the last years. Earlier, the control of the subsea wells was regarded as completely separate from the surface equipment. Now the subsea and surface equipment are considered together as an integrated production system.

The paper describe the design philosophies and standards used based on experience from 13 electro-hydraulic subsea production systems. The pros. and cons. for integrated and stand alone systems are explained in detail highlighting the preference for integrated systems.

Field experience of four integrated subsea control systems are described in detail.

## BACKGROUND

Kongsberg Offshore a.s. (KOS) has more then 15 years of experience in the development and delivery of subsea production systems. The first subsea control system development termed SWACS started in 1982 for AGIP as an autonomous subsea control system. It was installed on the Luna field in 1987 and is still operating. (See Subsea Control and Data Acquisition Conference 29-30 April 1992.) In 1988 the present generation of subsea control systems was put into production, based on experience from the SWACS system. KOS has delivered or in the process of manufacturing a total of 13 control systems with 101 control modules based on this technology. During 1994-95, the subsea part of the system has been considerably upgraded, in both functionality and hardware. The first delivery of such systems will be in the middle of 1996.

The topside (production platform or vessel) part of the subsea control system has over passed the years through several phases of development, from direct hydraulic control as delivered to North East Frigg in 1980 to the latest solutions with subsea control fully integrated into the standard topside Distributed Control System (DCS).

## TOPSIDE SUBSEA CONTROL UNIT (definition)

The topside subsea control system is referred to in this paper as the Subsea Control Unit (SCU, NORSOK definition). The functional requirements of the SCU are to facilitate control and to monitor and safe guard subsea equipment such as wells, manifolds and pipelines.

The special feature of the SCU is it's ability to control and monitor equipment which are not physically located on the topside production facility itself, but connected via subsea electro-hydraulic umbilicals, often several kilometres away. The SCU control via the umbilical is achieved by applying a dedicated communication protocol specifically designed for the purpose.

Traditionally, the SCU has been a stand alone system connected to the topside process control and shutdown systems via serial or hardwired interfaces. The SCU has in many cases been equipped with a separate (local) operator station providing all control, monitoring and diagnostics features. Limited remote control and monitoring functions have been implemented in the platform control system operator station via the dedicated interface.

Over the last few years, the design philosophy of the SCU has changed from the typical stand alone system to a more integrated solution, where the SCU is an integral part of the topside control system. This has been realised by utilising hardware, software and operator communication interface of the same brands as those of the topside control system, thus just adding a new member/node to the family of topside control units.

This paper is focused mainly on the integrated solution, but also tries to make a comparison of the pros & cons related to the stand-alone and integrated SCU designs.

# REQUIREMENTS FOR INTEGRATION OF TOPSIDE AND SUBSEA CONTROL

## By Authorities, Regulations and Standards

Requirements for integration of subsea control with the topside control system are not affected by any regulations. However, the following standards quote strong recommendations for such integrated solutions:

- *Norsok Standard, Subsea Production Control Systems, U-CR-005, Rev. 1, Jan. 95*

  The Norsok Standards cover all aspects of the offshore oil and gas industries, and are developed and supported by all the major operating companies on the Norwegian Continental Shelf. The subsea part of the industry is covered by several standards. The standard for *Subsea Production Control Systems* covers the total control system, subsea and topside. The following is quoted from *section 5.1.5 Topside Systems:*

  > *"Subsea wells shall be monitored and controlled from standard SAS (platform Safety Automation System) operator stations. Temporary operator stations for testing, commissioning, programming and maintenance can be included.*
  >
  > *The operator station VDU pictures shall be designed to topside system standard and fully integrated.*
  >
  > *The platform installed subsea control system shall functionally be part of the platform SAS.*
  >
  > *The production control system application software should be integrated with the SAS software. Preferably from SAS the control and monitoring of subsea functions shall be as similar to control of topside located equipment as possible.*
  >
  > *Control of topside utility systems for subsea installation (HPU, SPCU and UPS) should be performed by standard SAS nodes."*

- *Proposed ISO Offshore Standard, Subsea Production Control Systems, ISO 13628-6 (only working draft, issued Oct. 95)*

  The proposed ISO Offshore Standard is based on converting the Norsok Standards to ISO Standards, therefore, the text quoted in *section 7.4* of ISO 13628-6, is in a sense requiring the same as above.

**By Operating Companies**

The following four subsea projects, recently executed by KOS, have been used to illustrate operating companies demands on integrated control solutions. Two of the systems are in operation and two are close to delivery.

- AGIP Thelma Project

  Agip had prepared very extensive specifications for the Thelma subsea control system. The topside part called ECS (Electronic Control System) is located at the Tiffany platform. The specifications of the ECS defines clear requirements to utilise the same hardware and be designed according to the same functional requirements as for the Tiffany "Bailey" Distributed Control System (DCS) for topside production control. Under normal operation Thelma shall be controlled from the operator station in the Central Control room.

- ARCO Gawain Project

  Two options where defined in the Arco specification for the topside subsea control system, located on the Arco Thames Platform. One option was to modify/expand an existing subsea control system. The other option, which was selected, was full integration into the Thames Platform "Fisher-Rosemount" DCS for topside production control.

- BP Magnus SWIFT Project

  The detail specification for the subsea control of the Magnus SWIFT project was developed during a BP defined "FEED Phase." This specification defines clear requirements to full integration of subsea control into the Magnus Platform "Honeywell" DCS for topside production control.

  BP was running a separate project, in parallel with the SWIFT project, for implementation of Honeywell TDC 3000, as the new topside control system on Magnus.

- Statoil Norne Project

  In line with the above references, the Norne Project also specified clear demands to standardise on "Simrad Albatross" for all topside control systems, including subsea control. The difference between Norne and the other referred projects is that implementation of the application programs in the Simrad system was done in-house by KOS personnel with experience of programming Simrad Albatross equipment.

## COMPARISON OF STAND-ALONE VERSUS INTEGRATED SCU DESIGN

### Stand-alone SCU Solutions

Design requirements for the SCU are normally defined by the operating company and in many cases referred to as "the subsea communications unit," being a "black box," alien to the family of standard platform hardware and software systems.

The fact that the equipment is to be controlled remotely from the platform and even installed subsea, often seems to create the need for a non-standard solution and thus the requirement for a stand-alone system.

Benefits of the Stand-alone SCU

- The vendor can standardise on equipment hardware and software and thus lower his cost
- Free selection of hardware, software and graphical interface
- Simplified project execution with respect to implementation of application programs
- Limited restrictions/detail requirements by company on software engineering
- Design and detail engineering carried out by the vendor without the interruption of changes in design requirements for the topside DCS
- Can be fully function tested as a stand-alone unit.

Disadvantages of the Stand-alone SCU

- Additional level of computers between the DCS and SCM slowing down the communication
- Slow and "static" link between the DCS and SCU is normally the only feasible solution, e.g. MODBUS protocol is used
- Hardware and software different from the topside DCS
- Different graphical operator interface may result in incorrect decision making in critical/emergency situations
- Need for implementation of interface software and function logic in both topside control system and stand-alone SCU
- Requirement for duplication of database, local in the SCU and in the topside DCS
- Definition of local and remote operator functions
- Need for special operator training on the stand-alone SCU
- Need for suppliers service/maintenance support
- Requirements for separate detailed user and maintenance documentation
- Complicated and possible risky operation during start-up and maintenance, where alarm- and shutdown overrides are activated.

Generally we can say that our experience with stand-alone SCU is positive in the phase of design and engineering but not as positive when it comes to installation, test, commissioning and in particular operation and maintenance. After installation offshore the weakness of a solution with different hardware, software and graphical presentation on VDU's becomes more conspicuous. Frequently this leads to significant changes both in the SCU and in the topside DCS functional logic and operator interface. By experience, the communication and logic interface between SCU and DCS has caused considerable technical problems and increased cost.

**Integrated SCU Solutions**

In recent projects the benefits of an integrated topside subsea control solution have become more obvious. The operation of subsea wells is a natural and integrated part of the topside process systems, and all recognised suppliers of DCS today have the ability to meet all technological requirements for such solutions. An integrated solution means that a processing unit within the system of topside control units is dedicated to the task of subsea control and monitoring. The unit is connected to the common data bus as one of a number of distributed units forming the total topside control system. A common operator interface and graphical presentation is implemented on the operating stations for topside process and subsea control.

The SCU is equipped with standard DCS hardware electronic modules. The application program for subsea controls is implemented within the frames of the project/topside standard software. The interface to the Subsea Control Modules (SCM) is typically realised by off-the-shelf serial interface (RS 232C) modules programmed to handle a customised subsea communications protocol.

In recent projects, subsea production system contracts have been run in parallel or slightly behind their corresponding topside control system contracts, simplifying the process of adopting the philosophies and standards of the topside project. This, however, calls for tight co-ordination (interface) between the subsea and the topside project.

Benefits of Integrated SCU for one platform

- Common software, hardware components and data network
- Common operator interface and graphical presentation
- Increased safety through standardised shutdown handling
- Consistent process database, alarm- / event handling and reporting
- Simplified installation, testing and commissioning
- Maintenance of software and hardware by topside personnel
- Limited need for supplier's system support
- Consistent alarm- and shutdown override operation.

Disadvantages of Integrated SCU

- Integration with older platform process system may be limited due to poor availability of equipment and engineering documentation
- Land based test or pre-commissioning can only be performed with topside control equipment and graphical interface used as simulator
- Increased engineering effort and document interface between subsea contractor and DCS contractor.

# EXPERIENCE FROM RECENT PROJECTS

For most aspects of engineering, delivery and offshore installation, KOS's experience with integrated subsea control systems is positive. The experience of implementing conformity with project standards tends to result in less complicated and more "streamlined" procedures for hook-up, commissioning and start-up.

The following is a summary of experience from the four refereed projects:

**AGIP Thelma Project**

The Thelma deliverables are installed as an extension of the subsea field Toni (supplied by others).

Agip specifications are detailed and prepared by engineers with practical experience from the installed systems. The specifications are focused on high degree of standardisation / integration related to Tiffany / Toni, which resulted in:

- Re-use of project documentation from sub-contractor, slightly modified
- Re-use of test equipment and test procedures from sub-contractors
- High degree of re-use of static picture elements from the Toni node.

Thelma existing subsea field can be operated in the same manner as Toni. Display lay-outs and hierarchy are the same as for Toni, which resulted in:

- Reduced engineering discussions
- Reduced possibilities for erroneous operation
- Reduced training needs
- Reduced need for operator and maintenance manuals
- Reduced operational stress due to common operator picture layout

Thelma has the same electronic cabinets as for the existing units and shared/common spares. The existing operator stations have been used.

Software and Bailey firmware are the same as Tiffany / Toni.

**ARCO Gawain Project**

At the start of the Gawain project the contract was based on expansion of the present stand-alone "Orwell" subsea control system with the three Gawain wells. One option was that KOS should deliver a new stand-alone system. But after some informal meetings with the vendor for the Thames topside control system "Fisher-Rosemount," a proposal for an integrated solution was presented to ARCO.

Once the decision for an integrated solution was taken, the project went very smoothly, and the experience gained from Gawain is generally the same as from the Thelma project.

**BP Magnus SWIFT Project**

The decision to perform subsea control directly from the Magnus Platform Process Control System (PCS) as an integrated solution, was taken early during the concept phase (defined by BP as the 'FEED' phase). The Honeywell PCS concept was developed in a separate project, called 'MAGIC.' The interface and KOS deliverables were thus defined to be detail documentation of the KOS communication protocol and a functional data-specification. KOS did not have a formal, direct contact with Honeywell, hence problems that Honeywell had, where sorted out between BP and Honeywell.

The Honeywell-equipment had some limitations regarding capacity for tag numbers and communication lines. These were identified by BP/KOS early in project. The implementation was configured to handle these limitations and give an integrated system.

Points of interest:

- Well defined interface documentation
- Minimum requirements for follow-up by KOS
- Little or no requirements for KOS documentation to be updated due to changes in Honeywell equipment configuration
- Relative late verification of performance of the integrated subsea system.

**Statoil Norne Project**

The decision to integrate subsea control as a part of the Simrad Unified Automation System (UAS) on the Norne Production Vessel, was taken before KOS entered into the contract. KOS purchased a separate control node from Simrad, and has been totally responsible for application programming of the topside system, based on a Norne standard application manual, prepared by Simrad.

KOS had already experience with the Simrad Control Systems from two earlier subsea projects with Statoil and Conoco, the Statfjord Satellite Project and the Heidrun Project, both carried out as integrated system solutions. Therefore, when the Norne project started, KOS had gained considerable experience, both with Statoil specifications and with the control system.

Points of interest:

- The total control of the topside system by KOS, application programming and test
- Standard HW, SW and operator interface for the total Norne project
- Less documentation work
- Any changes of standard application software, in a late stage of the project, had implications on several contractors, utilising the UAS.

## CONCLUSION

On the basis of past experience and requirements defined by standards and operating companies, KOS will fully support and further develop the integrated SCU and DCS solution. This will lead to prime DCS vendors' implementation of the SCU functionality as a standard system feature, in-line with other topside process applications, as separation control, gas compression control, etc.

Effort and development from KOS will primarily take place on further standardisation of its own products and equipment, both subsea and topside. Effort will also be invested in increasing quality and reducing volume of the engineering and interface documentation.

Conversely, KOS has delivered stand-alone solutions, and will continue to offer such systems for projects where required by the customer, or for retrofits on older installations with no centralised DCS for topside production control.

A further reason for continuing the development of stand-alone solutions at KOS, is the ongoing general change of Process Control Systems technology towards "open system architecture" and internationally standardised software and communication protocols. Such changes will enable the possibility of interfacing systems from different manufactures. KOS can offer the market stand-alone solutions that fulfil these technology requirements. This is achievable through optimal selection of already available technology that fits with selected topside system. Such solutions will have the ability to incorporate the advantages of integrated solutions with shared operating facilities, common database and standardised communication interfaces.

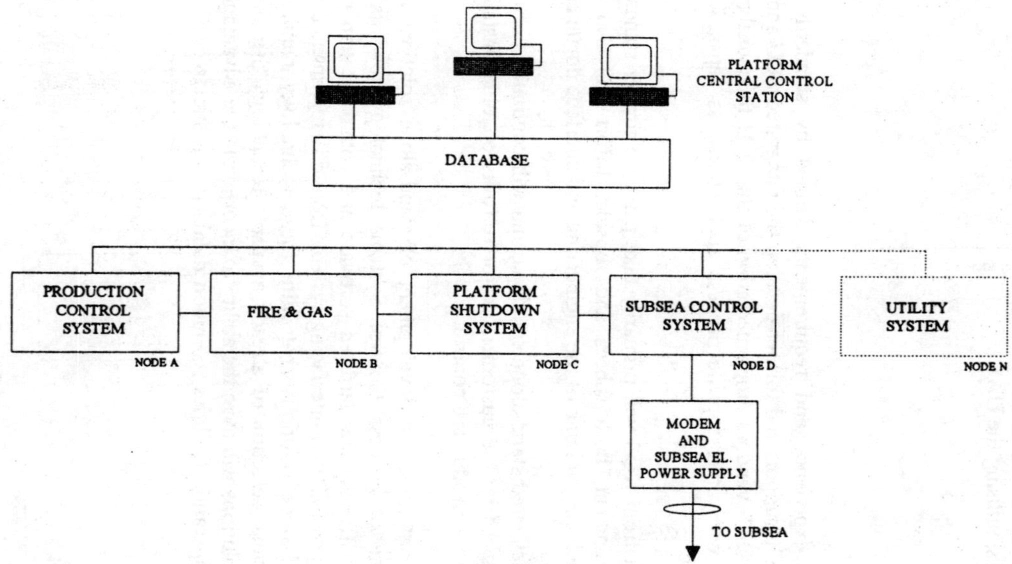


Figure 1. Integrated Subsea Control Units

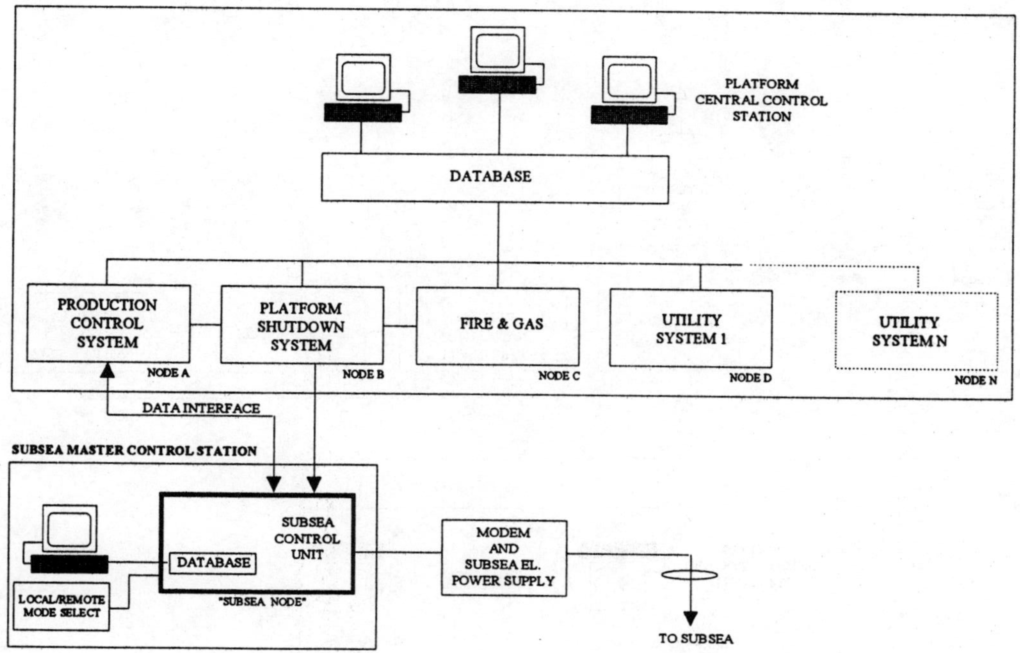


Figure 2. Stand Alone Subsea Control Units

# )eployment instrumentation and control system or a gravity base tank

NORMAN
ubsea Offshore Limited


## SYNOPSIS

The BP Harding field utilises a concrete Gravity Base Tank (GBT) combined with a specially designed steel jackup to provide the production platform. This paper is concerned with the GBT and in particular with the deployment instrumentation and control system. The paper therefore describes the GBT, its function and design, the requirements of the deployment control system, its actual implementation and finally how this worked in conjunction with ROV intervention to facilitate the installation process.

## 1 INTRODUCTION

The BP Harding Gravity Base Tank (GBT) is a 'T' shaped structure constructed from reinforced and pre-stressed concrete that provides a foundation and oil storage facility for the TPG500 production, drilling and quarters (PDQ) jack-up platform located upon it. In June 1995 the GBT was successfully installed in the Central North Sea, approximately 320km north-east of Aberdeen in 109m of water.

The deployment instrumentation and control system was designed and built by SubSea Offshore Ltd. in Aberdeen and subsequently installed on the GBT. This work was carried out under a subcontract from Taylor Woodrow Management and Engineering Ltd. (TAYMEL).

The main structure was designed by Taywood Engineering Ltd (TEL), for construction by the Costain Taylor Woodrow Joint Venture in a dry dock at the Hunterston Marine Construction Yard in Ayrshire. The installation control and monitoring systems built into the GBT also had to be integrated with temporary systems and equipment provided by Norwegian Contractors who were contracted by BP Exploration and it's partners Repsol and Ranger, to carry out the offshore installation operations.

## 2 GBT CONCEPT AND STRUCTURE

Before describing the GBT installation control and monitoring systems it is necessary to give some background to the GBT itself. Reference should be made to figures 2 and 3. For structural reasons the tank is internally composed of 80 cells, each approximately 9m square and 18m high. Four of the cells are penetrated by the drilling conductors and are not used for oil storage. The remaining 76 cells are variously interconnected to provide three main tank sub-sections, which, during operation, are filled with crude oil floating on a cushion of sea water in an 'open storage system'. The overall dimensions of the 'T' shaped structure are approximately 114m wide by 122m long and 20m from the base to the roof of the main tank. The height to the top of the three towers that support the jack-up legs is 34m. Two metre deep steel skirts protrude from the base of the tank and penetrate the sea bed to help stabilise the foundation. Twenty-eight open topped cells surround the structure and extend to 9.6m above base slab level. These cells provided additional buoyancy whilst floating the GBT out of dry dock and were subsequently loaded with approximately 15,000 tonnes of stone ballast to add bottom weight. As a further aid to float out of the dry dock the GBT was constructed from a special concrete, using a man-made lightweight aggregate, some 20% less dense than normal concrete.

On completion of construction the dock was flooded and the GBT was floated into the Clyde Estuary for final trials prior to sea tow and installation. To assure the required hydrodynamic stability during tow to field and installation on the sea bed the tank was sub-divided at low level into 17 watertight compartments for the segregation of ballast water. At float-out from dry dock the GBT and the external buoyancy cells had to be empty. Following the ballasting trials, moored off Largs, the GBT contained approximately 2m depth (12,500 tonnes) of ballast water, evenly distributed between compartments and all the solid ballast had been added.

At the installation location, it was necessary to add a further 48,000 tonnes of sea water to bring the structure to a neutrally buoyant condition. The descent to the sea bed was achieved by the careful addition of further relatively small quantities of sea water to overcome the restraining force of three strings of large floats attached to strong points on the main roof at the ends of the structure. Other than the float strings, there was no other surface penetrating body to provide stabilising forces during the descent to the sea bed. For this reason accurate and reliable control and monitoring of the ballasting process and the trim was absolutely essential. In addition, during all stages of the installation, the difference in ballast water head across any internal compartment wall had to be limited for structural reasons to less than 3m and air pressures throughout the GBT tank and towers had to be monitored.

To accomplish the installation, ballast water was admitted through an arrangement of 20 manually remote controlled, hydraulically operated valves. Air pressure and water level at several positions within the tank, together with inclination and altitude (i.e. height above sea bed) of the structure, had to be measured and displayed to the operator on board a dynamically positioned surface vessel (SSOL's MV Discovery). Transmission of hydraulic fluid and electrical power and control signals was via an umbilical provided by Norwegian Contractors.

After safe landing on the sea bed the ballasting process was continued to complete the flooding of the structure and then to vent the air trapped in the tank and towers by ROV operation of vent valves and remote operation of ballast valves. Again, this process had to be carefully monitored and controlled to avoid excessive air or ballast water pressure differentials within the structure and to ensure an even driving of the skirts into the sea bed.

## 3 INSTALLATION CONTROL SYSTEM REQUIREMENTS

Given the installation concept described above TAYMEL developed a requirements specification for the control and monitoring systems. Because the ballasting and monitoring systems would be permanently installed into the GBT structure and used only once, minimum capital cost was desirable. However, high reliability would be essential because failure of the ballasting control or monitoring equipment could jeopardise the installation of the £50m GBT and thereby the £400m Harding Field development itself.

It was the design intent that the deployment procedure should be capable of completion in the event of any single failure of an active component. The system design therefore incorporated redundancy, diversity and separation of equipment. Since the twenty ballast valves and their hydraulic actuators were not duplicated their operation was backed-up by the capability to make direct ROV intervention to open or close valves. An independent fault tree analysis was carried out on the final system design to justify its acceptability for use.

The required control and monitoring system functions were as follows:

- Operation of the ballast valves via a push-button console enabling each valve to be opened or closed by the operator on board the installation vessel. Each valve to have status feedback to the operator to indicate if the valve is open or closed. Each valve actuator solenoid valve to have two independent coils either of which would operate the hydraulic pilot valve.

- Absolute air pressure to be monitored at two positions within each tank section, the drilling compartment and each tower.

- Water level within each ballast water compartment and each tower to be monitored by two transmitters, one using a differential pressure measurement and the other based on a sonar device.

- The height of the base of the tank from the sea bed to be measured at three positions on the GBT (at each end of the 'T' shaped structure).

- Inclination of the GBT in both the x and y axis measured by duplicate instruments mounted firmly to the structure.

As a general principle the use of proven "off the shelf" equipment was favoured since, in the time available, this would be more likely to be reliable and would have greater security of delivery than one-off, purpose developed equipment.

The required data acquisition, processing and display system functions were the following:

- Duplicate signal processing and communications units, installed in pressure vessels (Control Pods) suitable for sub sea use and located on the main roof of the GBT.

- A duplicate, fully redundant PC based operator interface system located on the installation vessel to indicate by means of VDU's all data and alarms and to carry out all calculations required for the deployment indications. A security system to be included to prevent unauthorised change of programs or data.

- A serial link to the surface interface with the readout equipment and valve controls and also to supply all raw data to the Installation Contractor's monitoring system via an RS232 serial link. Each link between the GBT and surface equipment to be through a single screened pair in the umbilical connection cable.

The Installation contractor also had a portable PC terminal that could be linked into the system at the umbilical termination on the GBT roof should it be required to interrogate the ballast level and inclinometer instrumentation during the sea tow - e.g. in case of a suspected leak.

The values returned from the monitoring instruments were displayed to the operator via purpose built VDU displays. The differential head across each compartment wall was calculated and graphically displayed, with the system giving an audible and visual alarm if a preset value was exceeded. The compartment level was also corrected to display the level at the centre of each compartment (since the measurement point was not at the centre point of the compartments). Differential air pressure between tanks was calculated, with an audible alarm given if a preset value was exceeded.

### 3.1 Site Installation

Other important features for the instrumentation and control systems were ruggedness and ease of installation. Because of the very tight construction programme (17 months from first permanent concrete to float out) equipment had to be built into the GBT structure in close proximity to heavy civil works concerning the completion of the towers and the tensioning and grouting of pre-stressing cables.

A key feature was the prefabrication of all system components and cabling. These were pre-commissioned at SSOL's Aberdeen Base prior to shipping to site, see later.

### 3.2 Hydraulics

The hydraulic oil was supplied by umbilical from the surface hydraulic power unit via a pressure stabilising system at the Installation Contractor's Mechanical Equipment Module (MEM) situated in cell C8. Vented oil was returned to the surface.

There were facilities to allow any of the ballast valves to be directly operated by ROV, see later.

## 4 ICS IMPLEMENTATION

### 4.1 Overview

The system comprises topside hardware and subsea hardware. The subsea hardware has been designed using standard techniques developed by Subsea Offshore for Remotely Operated Vehicles and Intervention Systems. Therefore the designs are somewhat different to conventional plant control and instrumentation but utilise some special electronic hardware and packaging. The latter is particularly manifested in the underwater pressure vessels or pods.

Underwater, two pressure vessels, see fig. 4, were installed on the GBT roof, adjacent to tower three. The actual physical pod design is that of the Examiner series of ROVs. Each of these pods contain duplicate electronic hardware. The system controls solenoid valves as part of the ballast system and reads water level sensors positioned within various compartments. There are also absolute pressure sensors, sonar sensors and inclinometers as part of the system.

The sensors are not fully duplicated but are connected to the multiplexers such that there is always a measurement of level and pressure for each compartment even if a complete multiplexer system malfunctions.

Topside there are two complete control systems housed in one free standing console. Each system consists of a PC in Euro format connected to an SSOL Control multiplexer. The control panel itself is composed of discrete switches in a T shaped mimic, directly connected to the Control Multiplexer. The readbacks are displayed graphically, being generated by the PCs. Two monitors are able to display the same range of information on line. There are various displays of water level, valve status and calculated differential pressure across compartment walls for example.

In addition to this hardware, the system comprises underwater cabling to valves and sensors. In all, there is over 10km of cable involved and over three hundred connector and penetrator assemblies.

## 4.2 Control Multiplexer

The heart of the control system is the Control Multiplexer. This is a single board multiplexer, in this case in eurocard format, which is placed in each control pod. There is a corresponding board for each multiplexer at the surface console. The communication system is therefore fully duplicated.

The Subsea Offshore Control Multiplexer is a single board containing all the electronic circuitry to drive several kilometres of twisted pair cable and provide the user with analogue, parallel digital and serial digital interfaces. In this case the umbilical length was only 500m. The basic specification of the multiplexer is as follows :

- Full Duplex Transmission over 5.5 kilometres of twisted pair.
- 8 channel 12 bit -10 to +10 volt analogue/digital interface.
- 8 channel 12 bit -10 to +10 volt digital/analogue interface.
- 48 digital bits configurable as inputs or outputs.
- Full duplex RS232 serial channel with 38.4 kbaud maximum speed.
- Additional 9600 baud RS232 port for control/debug purposes.
- Expansion port for additional digital and analogue interfaces.

- Expansion port increases digital I/O by 16 bits when free.

- Board size of 119 x 129 mm or as standard eurocard (100x160)

- Highly secure data transmission for control channels.

- Transmission conforms to international telecom standards.

The SubSea Offshore Control Multiplexer (Mux) provides the communication along long lengths of cable between two electronic interface stations. These stations typically provide control, display, sensing and drive electronics normally associated with control of remote vehicles and systems.

The ability to transfer data over a twisted copper pair at high speed and with high data security was paramount in the design concept, together with a flexible system configuration method.

All the electronics to interface with the cable, provide digital and analogue interfaces, sequencing, control of expansion boards, and serial channel interfacing, are contained on the single PCB .

The method of transmission of data is taken from the world of telecommunication over land lines. The reason this approach was taken is that this method has been subjected to many tests and trials and data is available on maximum cable lengths, error rates and noise levels. It is an internationally recognised method of data (and speech) transmission.

At each end of the cable, there is an Echo-Cancelling Interface Transceiver. This provides two 64 Kbit channels and a 16 Kbit channel in both directions, in full duplex mode - thus providing an ISDN Basic access data transmission. One of the channels is used to transmit the digital, analogue and expansion board data in an HDLC format, and the other channel provides for the serial interface link for asynchronous baud rates up to 38.4 Kbaud. The bandwidth of the signal is only 80 kHz which permits transmission over standard telephone wires up to 5.5 kilometres in length. The method is also far less likely to interfere with others signals (e.g. video) within the umbilical than other methods of coding.

The multiplexer was connected into a mother board or backplane specially designed for this application. Along with the mux were two valve driver boards and two input boards. Each valve driver had the capability to operate 12 valves in a bi-polar mode. In this case the coils were switched at 24Volts and only consumed 9 Watts. There were twenty valves each with duplicated solenoids. Again, one mux system controlled the 'A' coils, the second controlled the 'B' coils. All the valves controlled actuators-valves comprising the GBT ballast system. All actuators had limit switches fitted so that valve status could be reported to the surface. The actual control was open loop with time limits imposed from the surface console.

The two input boards were designed for either current loop or voltage input. This was switchable on the board. Each board had a capacity of 16 inputs. The pressure sensor outputs (see below) were current loop with the acoustic sensors being voltage sources.

The mother board was a four layer board measuring 295mm by 210mm. Although only two systems were built, it was decided that the amount of connections between boards and also to the external equipment via penetrators justified such a design approach. Included in the mother board was some low pass filtering designed to take account of the acoustic sensors' effect on power supplies and of course sockets to enable the pod electrical penetrations to be connected to the electronics rack. This combination enabled time spent on testing and assembly into the pressure vessels to be minimised.

Once commissioning was complete a vacuum of approximately one quarter bar was pulled on each pod. This was monitored on line by a sensor in the pod and displayed at the surface console. In fact, once the pods left the Aberdeen base to be delivered to site at Hunterston, the pods never required opening and the vacuum remained stable throughout the period to installation day some five months later. Similarly, the electronic hardware by and large was continuously energised and well burned in come the big day.

Besides vacuum status, the pods were also monitored for temperature and water ingress.

## 4.3 Sensors

A number of types of measurement were necessary in order to control the installation process. These consisted of level measurements within the GBT compartments and towers, pressure measurements in the towers and certain compartments, inclination measurements of the GBT in two axes, depth measurements and valve position indication as discussed above. In order to achieve this the following sensors were used.

Level measurements were obtained using two types of transducer. The first method was to use differential pressure sensors in stainless steel housings fitted with standard underwater electrical connectors. The sensors were 4 to 20mA current loop and were connected to the first multiplexer system. The accuracies were better than 0.4% which, when combined with other considerations gave an overall accuracy of 85mm of water depth. One pressure port of each transducer was fitted with hydraulic hose and the other end of the hose was installed, open ended, in the roof of the compartment thus giving a reference pressure. The second port was open to ambient pressure via a filter and with each sensor fitted to the compartment floor, a measurement of depth was obtained.

In all twenty transducers were fitted, one in each of the three towers and the seventeen compartments. Alongside these were fitted a corresponding sonar device which was an echo sounder similar to those commonly fitted to ROVs. In this case, however, the echo sounders were actually looking upwards at the air water interface. They operated at a frequency of 500kHz with a range of 2 to 100m. In fact the internal height of the cells was approximately 18m. Their accuracy was better than 1%. The sensors had voltage outputs and these were connected to the second multiplexer. Thus, if either system suffered a major failure then there was a measure of water level in each compartment and the ability to control the ballast system from the one multiplexer system. Each echo sounder had a beam width of 4 degrees. Since in theory the GBT could be inclined at greater angles, in order to ensure stable measurements, the sensors were mounted on gimbals. Although of basic design, this enabled the echo sounders to be always vertical. In practice the readings obtained within each compartment were very solid and the sensors performed extremely well.

Absolute pressure measurements were made in the three towers and six of the compartments using fourteen sensors. These were similar in construction to the differential type, except for the differential feature of course. They were mounted inside their respective compartments just inside the ten inch penetrator assemblies at the top of the cells. All the tower sensors were fully duplicated and utilised both multiplexer systems.

Inclination measurements were made by installing a two axis inclinometer in each of the underwater pods which were mounted on the GBT roof. The inclinometer display proved to be very useful in practice. Mechanical means were provided for alignment purposes. The axes were defined with reference to the centres of the towers. The sensor had a range of +/- 15 degrees (the GBT did not actually adopt angles greater than 5 degrees) with an accuracy of 1 minute of arc, and had voltage outputs. At Hunterston, the inclinometers was calibrated using the services of the on-site surveyors.

Altitude measurements were made at three points around the GBT structure using three echo sounders of the same type as used above for level measurement. These were conventionally mounted looking downwards, being attached to the GBT base slab. They were calibrated to indicate the distance of the bottom of the skirt from the sea bed during submergence. The echo sounders were cabled back to both multiplexers, but the system was not duplicated. One initial problem did manifest itself with the bracket design which allowed a small pocket of air to be trapped on the transducer face. This was enough to impede the transducer causing misleading results. However, once identified, the cause was easily rectified, albeit requiring shallow water diving.

## 4.4 Underwater Cabling

The size of the GBT and the numerous sensors and their geographical distribution made it inevitable that a large amount of cables and connectors were required. In principle standard underwater techniques were used to implement the system. The general design goal was that no one cable fault could affect more than one sensor or valve function. With reference to figures 5 and 6, every device had its own individual cable assembly or assemblies which were routed all the way to one of the two control pods. In all there was over 10km of cable used and the number of penetrations into the pods exceeded one hundred and ten.

Since the pods are of a modular design, an additional penetrator ring was added per assembly over the normal ROV configuration, see fig. 4. In some cases, e.g. valve status readbacks, Splitconn connectors were used to double up on the number of cables connected via one penetration. Once into the pod the connections were direct to the mother board.

All horizontal cable runs on the concrete GBT roof were laid directly into cable trenches formed from rows of bricks with a sand bed between. As soon as all the cables in a section had been laid they were covered with a further layer of sand and pre-installed reinforcing bars were bent over to provide immediate protection against damage. Coils of part installed cables were protected in "Wendy Houses". Once all the cables in a section had been laid and successfully checked the relevant section was encased in concrete to provide permanent protection. Vertical cable runs were pulled into steel conduits. Considering the intense activity on the GBT at the time there was remarkably little cable damage and it was all satisfactorily repaired by in-situ splicing.

Tank and tower instrument packages were installed into ten inch diameter vertical steel penetration sleeves cast into the concrete roof panels, see fig. 6.. Strain wires to carry cables and DP dry leg sensing pipes were installed by roped access personnel. Two electrical penetrators were fitted to each one, designed to withstand external sea pressure. At each location, the two penetrators allowed sensors to be connected to multiplexer one or two depending upon the relevant grouping.

All the electrical penetrations and cable assemblies were manufactured in a factory environment and subsequently pressure tested including appropriate electrical checks at each stage of course. This obviated the need to carry out cable moulding on site as far as was at all possible. This method did prove effective in as much as although some problems were experienced during manufacture, they were put right without affecting the installation phase thus minimising cost and schedule impact. At the time of year, January and February on the Scottish west coast, the installation time was minimised since, apart from the problem of running so much cable and concreting it to the roof top, electrically every connection was a plug in job. In fairness to the on site cable moulding fraternity, successful terminations were being made at Hunterston on other parts of the permanent instrumentation system.

The control pods were connected to the surface via an umbilical supplied by Norwegian Contractors. This was terminated at one corner of the GBT roof and supplied 220Vac and an unscreened twisted wire pair to each pod. In addition, in the event that the umbilical became damaged, connections were supplied to the control pods from two intervention panels situated on the Mechanical Engineering Module (MEM) in cell C8. These panels enabled an ROV to connect to the pods such that power could be supplied from on-board ROV systems and its umbilical could provide a twisted wire pair for communication. At the surface, corresponding provisions were made between the ROV Control Van and the GBT Installation Control Centre.

## 4.5 Topside Hardware

One free standing console was used to house both topside control systems, see fig. 7. As described earlier each system centres around a Control Multiplexer which has a corresponding subsea board. Associated with each multiplexer topside is a single board PC, graphics card, I/O card, hard disc, power supply, monitor and sixteen function keypad.

A control panel provided one set of switches to operate all sub sea valves. The switches were hardwired directly to each of the control multiplexers. In this way a failure of a PC did not affect the ballast operation. For each valve, one switch was operated to enable the on valve state, the second being operated to release the open control and return to valve closure. Each button was illuminated from the feedback signal taken from the multiplexer.

All displays were graphically generated. The PC provided a timed override signal for each of the seventeen compartment ballast valves such that even when selected, a valve would be released to close after a time adjustable up to 1000sec. A single push button switch was also supplied to provide a 'general closure' facility such that all the above valves and the three main ballast manifold supply valves would close immediately this switch was operated.

The PCs were connected to the multiplexers via an RS232 serial link. Similarly, a serial link was provided to the Installation Contractor to supply all status information to his own monitoring system. The overall update rate was approximately six times per second.

All the software was written in 'C'. All information from either control system could be displayed on either monitor.

There are nine different displays which can be accessed by the users of the equipment. These are as follows :

**Page 1** The "Normal" display. Shows the water levels and pressures within each of the tanks and indicates the level differences in a format that replicates the actual layout of the GBT.

**Page 2** Shows the current status of the 20 valves. When valves are operated the colour of part of the valves changes to reflect the change that has been requested.

**Page 3** This shows the same information as on page 1 outwith level differences, but with the tanks ordered according to the ballast manifold to which they are connected.

**Page 4** Gives the operator information about the attitude of the GBT.

**Page 5** Information about water levels is displayed on this page in a tabular form.

**Page 6** The settings relating to valve timings are displayed on this page in a tabular form. If required the timings can be changed from this page.

**Page 7** The information from the water level sensors and inclinometers is used to display the calculate level differences between adjacent tanks.

**Page 8** If any alarms are present they are displayed on this page.

**Page 9** The sensor data from the three towers is summarised on this page along with sensor informati from within the two control pods.

Calculations - in addition to providing information about the level of water in the various compartments, the differential pressure transducer signals were used to calculate the water level differentials that could exist between compartments. This was one of the most important parameters to be displayed since excessive differentials could cause structural failure.

Alarms were provided for inclination, differential pressure and differential level, as well as water ingress and loss of vacuum inside the control pods.

## 5 ROV INTERVENTION

A number of features were integrated into the GBT design in order that backup services could be provided by ROV. This covered not only the installation control system itself, but also general GBT installation procedures.

Four ROV panels, see fig. 8, were fitted in cell B8 in order to provide override facilities to the solenoid valves controlling the actuator-valves making up the ballast system. The override was enabled using a standard ROV low torque control valve tool and in a similar way the actuators themselves could be operated directly using a high torque output valve tool.

As discussed above two panels were also fitted to the MEM in C8 to provide electrical intervention in the event of umbilical failure. Similarly, an hydraulic supply from the ROV system could be made available to the MEM systems.

In addition to these possible tasks the following were carried out by ROV:

- Seabed Survey, a video survey centred on the installation location was carried out to check for debris which may have been hazardous to the GBT installation.

- During the GBT descent, the ROV monitored the control umbilical catenary and the status of GBT touchdown.

- Once the GBT's position was accepted, the control umbilical was disconnected using ROV tooling.

- The ROV initiated removal of towing pins at four locations.

- Large surface buoyancy tanks were attached to the GBT which controlled the structure's descent. After installation, the buoyancy lines were cut by ROV.

- The MEM was then removed for future use. The ROV was used to remove clamps from the ballast lines, cut hydraulic lines and connect a crane hook to a sling arrangement in order to facilitate recovery.

- The ROV aided transponder recovery after they were no longer needed.

- Air vent valves fitted to panels at the tower roof tops were operated by the ROV.

- Grouting operations were carried out by lowering three grout lines as a bundle from the surface. At the line ends, female grout connections were staggered to ease the operation. The ROV connected these lines to grout risers and monitored primary grout vents during operations. Similarly, grout bundle catenaries were monitored.

- Finally, an installation survey was carried out with particular attention being paid to sea bed penetration and the possibility of debris around the structure.

## 6 CONCLUSIONS

This was a very successful project not only for the faultless way that the control systems performed during the actual installation, but also for the general project management and style set up by BP and all their subcontractors. By writing functional specifications and then allowing contractors to make timely decisions, the project proceeded well and with full co-operation between participants.

On the technical side there were some lessons to reflect on of course. These were essentially as follows :

- Whilst the echo sounders worked extremely well, it was easier to install and calibrate the differential pressure sensors. Therefore it may be better to use these exclusively, employing three instead of two sensors per cell and instigating a two out of three polling method. Although each echo sounder was individually tested prior to delivery to site, it was difficult to do a complete calibration because the cells were dry.

- There was some debate as to whether the d.p. sensors should have been installed in the bottom or the top of the cells. Three of these actually failed immediately prior to towing the GBT to the Harding location. One of the factors may have been their siting. However, with the cells flooded at this stage, it was not possible to recover the units to determine actual cause. The sensors and their cabling were replaced in an ad hoc manner to fix the problem.

- The cabling method on the GBT worked well. However, the strategy should be reviewed because some rationalisation may be possible.

- The topside displays possibly provided too much information for some operators. For example, the control system updates extremely quickly. At Hunterston, once the dock was flooded, sometimes there was about a 200mm wave of water rippling through inside the tanks due to the external weather. The depth sensors followed the wave very well and the displays relayed the information accordingly. However, it may have been better simply to integrate the values such that a constant reading was manifest.

- It would be possible to build in provision to recover the control pods at the end of installation for future use, in a similar way that the MEM was recovered.

## 7 ACKNOWLEDGEMENTS

The author would like to thank Taymel, and in particular Mike Griffiths, for his valuable contribution towards this paper and also for the general support and co-operation throughout a first class successful project.

Finally, thanks to BP Exploration for permission to publish this paper.

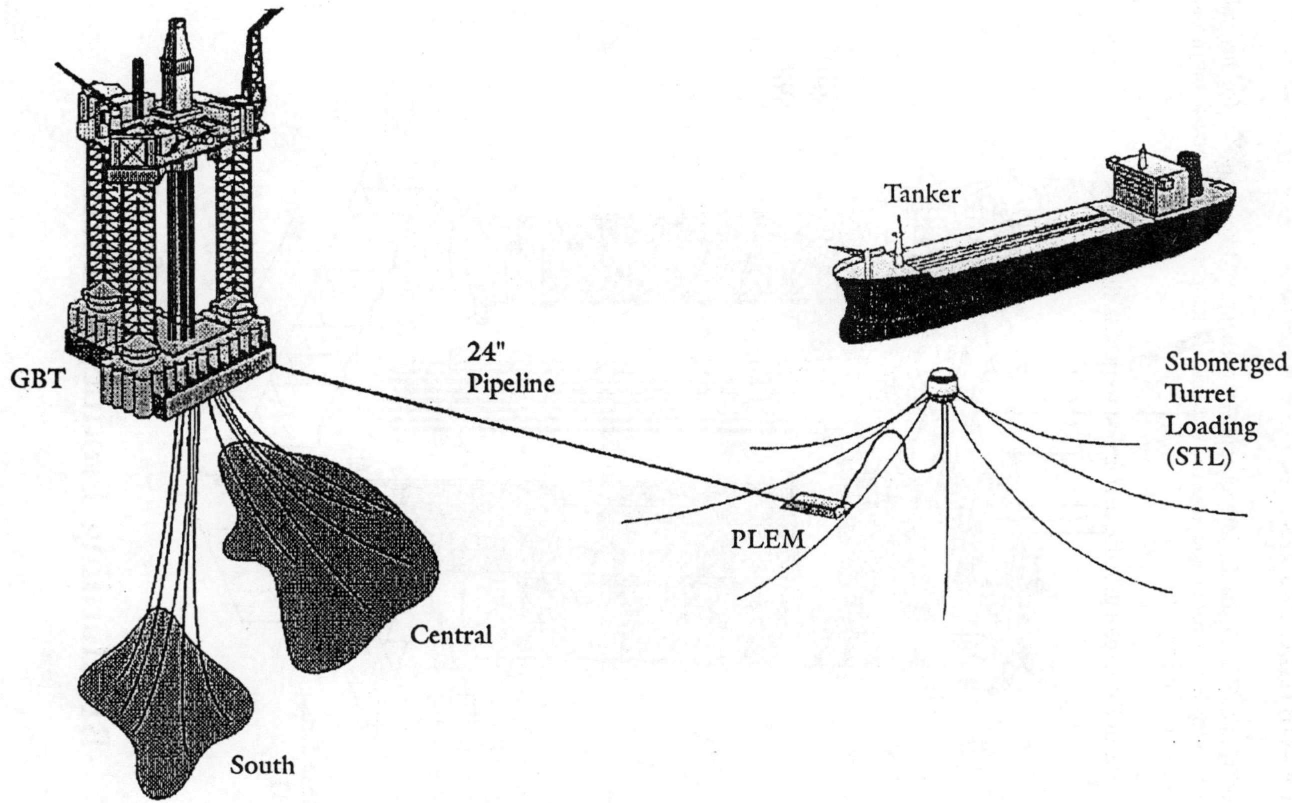


BP Harding Field Development

*Fig. 1*

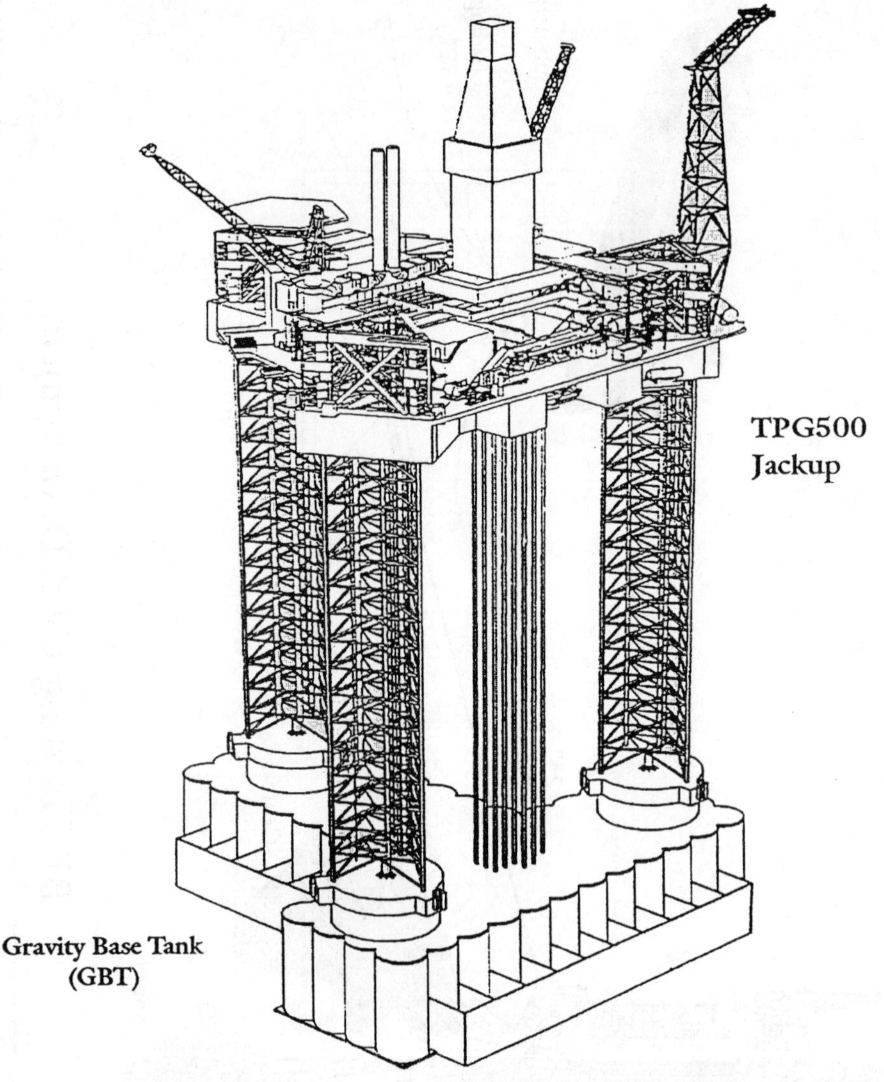


# BP Harding Production Platform

*Fig. 2*

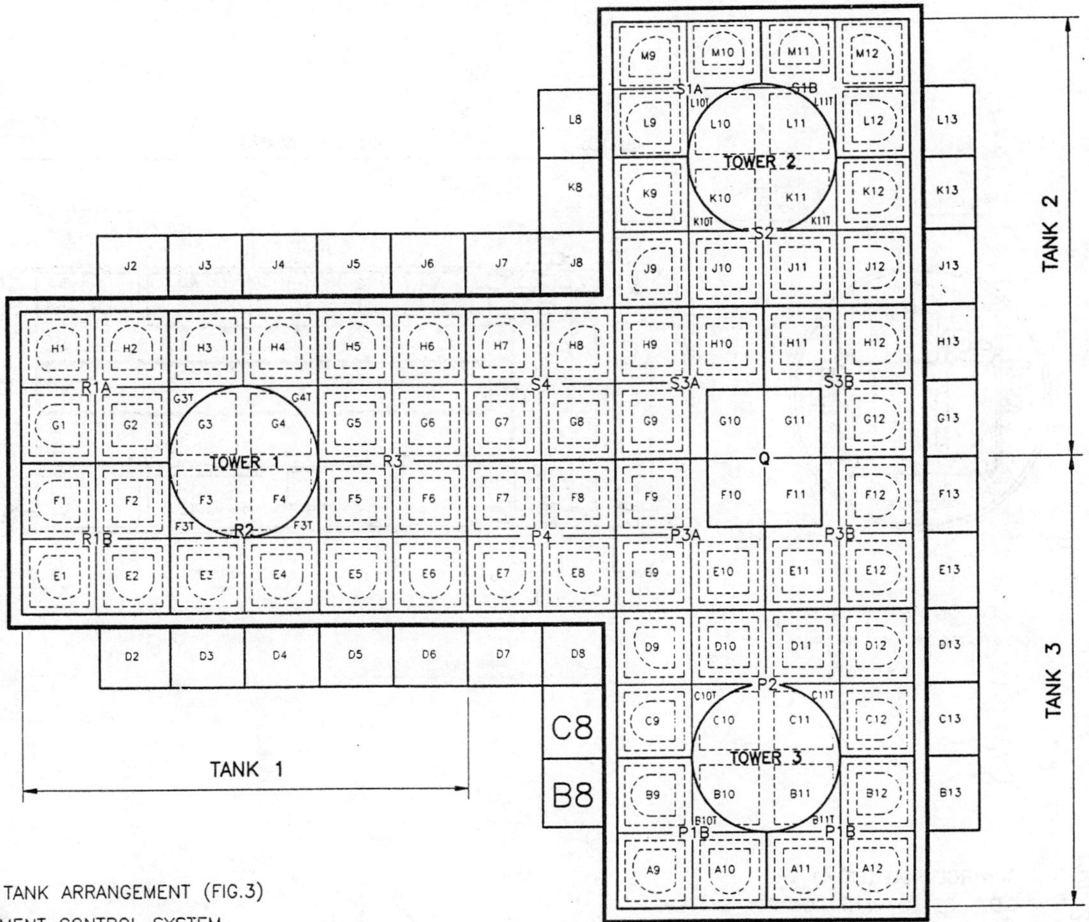


GRAVITY BASE TANK ARRANGEMENT (FIG.3)
G.B.T. DEPLOYMENT CONTROL SYSTEM.

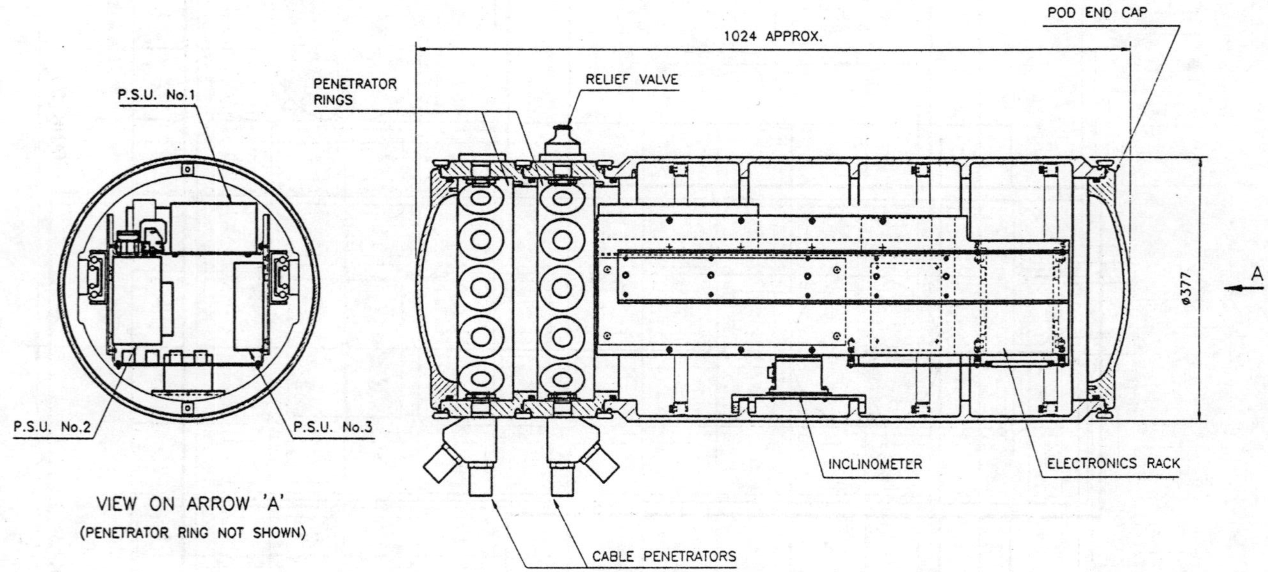


CONTROL POD (FIG.4)
G.B.T. DEPLOYMENT CONTROL SYSTEM.

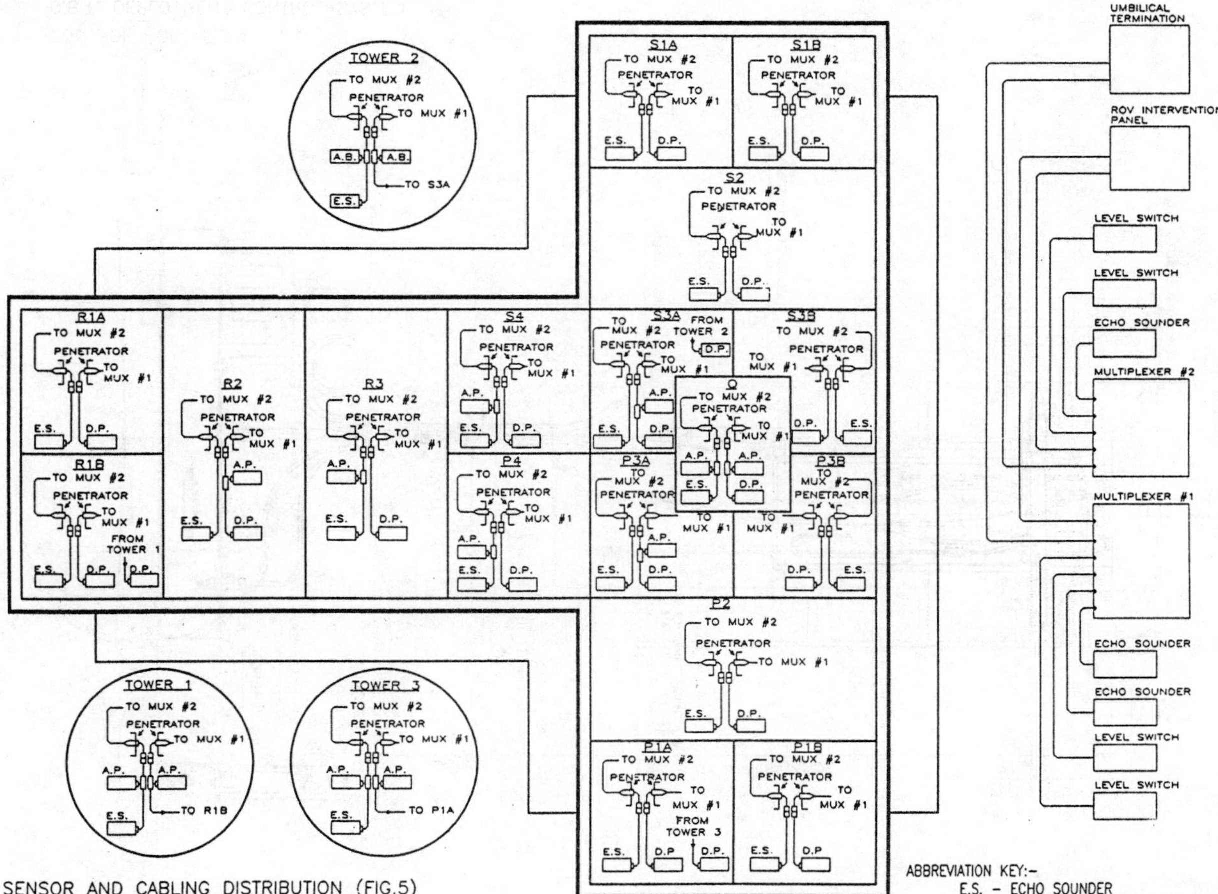


SENSOR AND CABLING DISTRIBUTION (FIG.5)
G.B.T. DEPLOYMENT CONTROL SYSTEM.

ABBREVIATION KEY:-
E.S. - ECHO SOUNDER
D.P. - DIFFERENTIAL PRESSURE TRANSDUCER
A.P. - ABSOLUTE PRESSURE TRANSDUCER

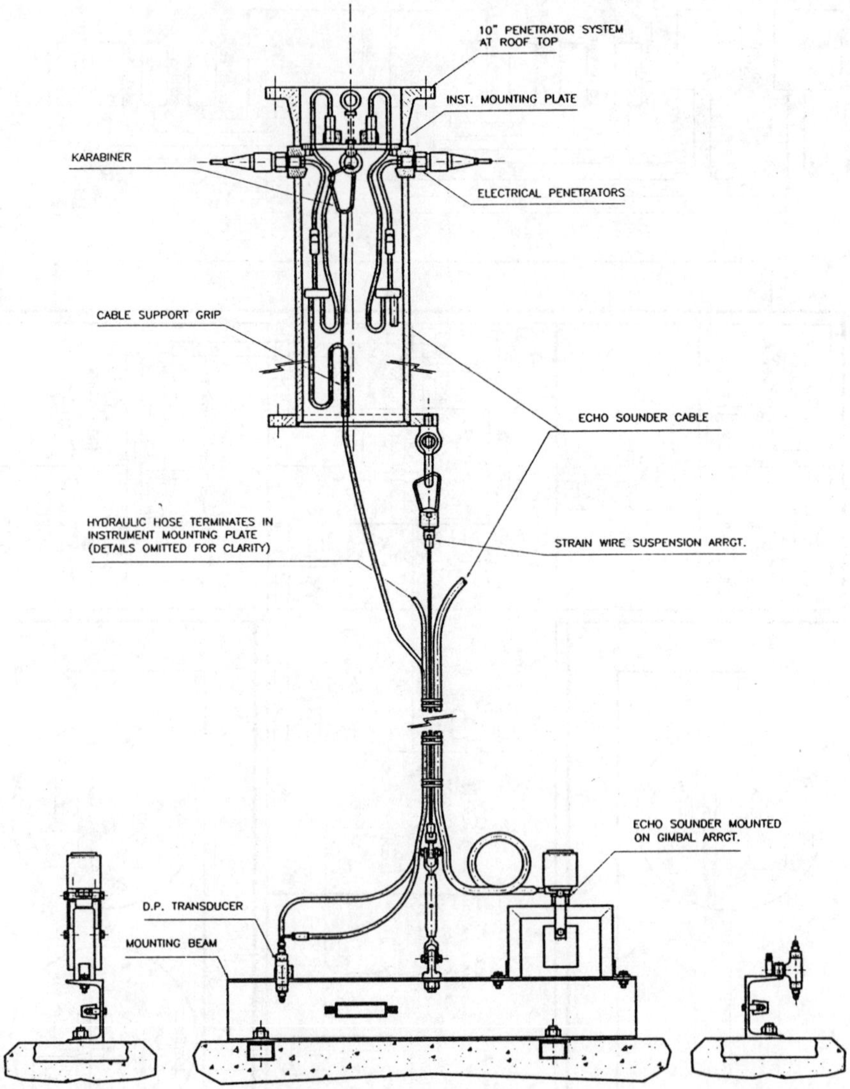


INSTRUMENT CABLE SUSPENSION ARRGT. (FIG.6)
G.B.T. DEPLOYMENT CONTROL SYSTEM.

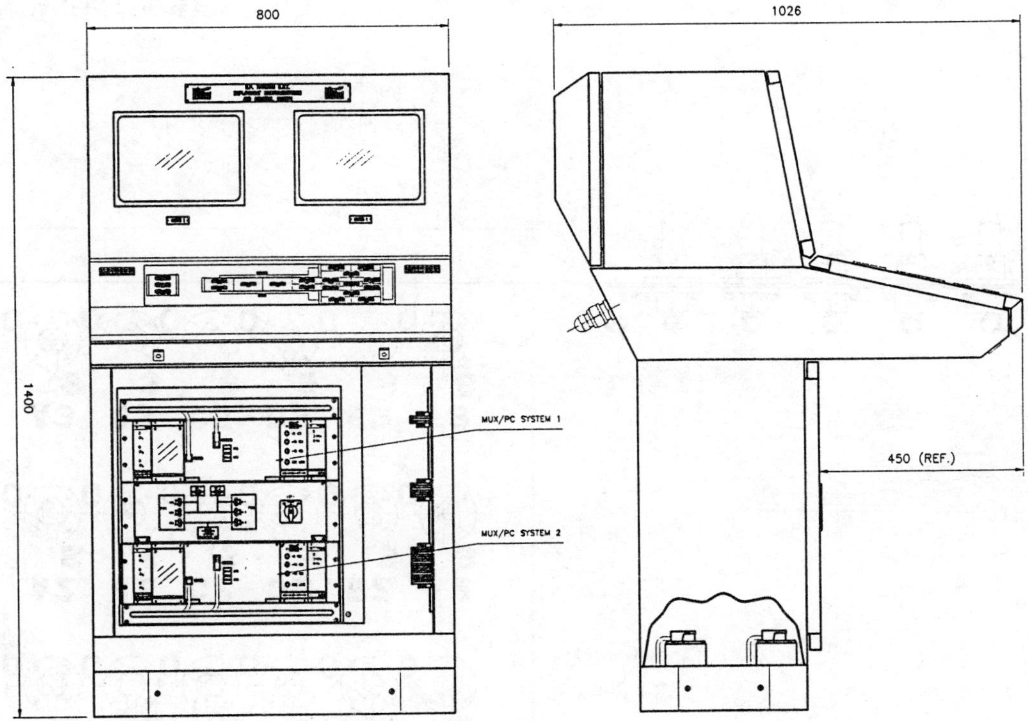


DOOR OMITTED FOR CLARITY

CONTROL CONSOLE (FIG.7)
G.B.T. DEPLOYMENT CONTROL SYSTEM.

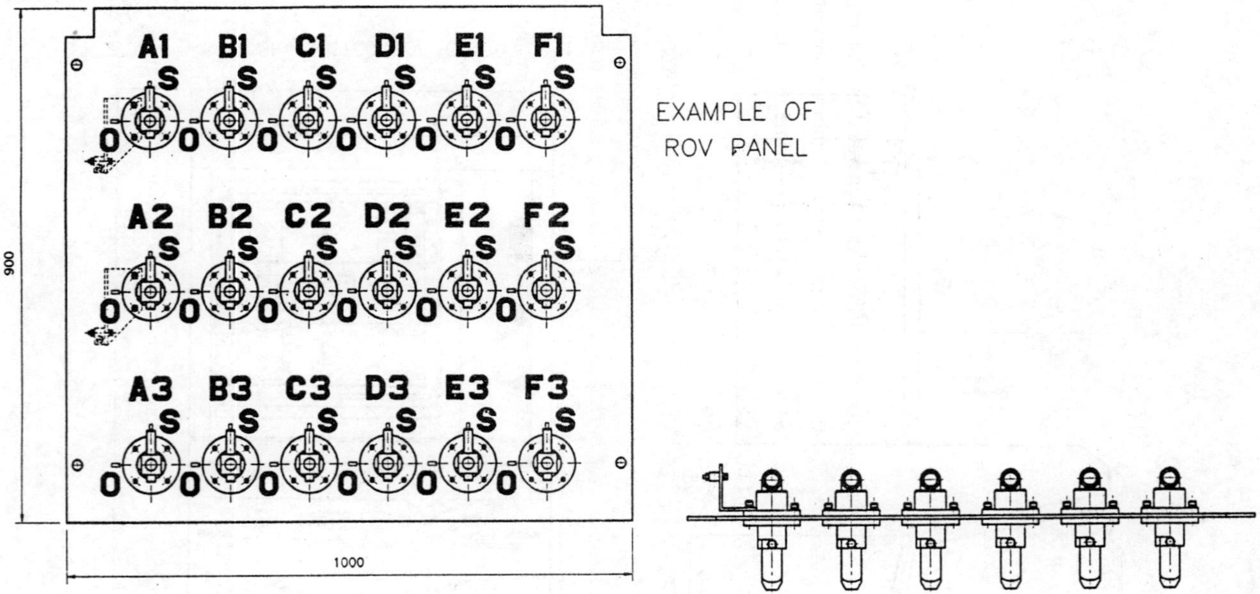


FRONT ELEVATION

PLAN

ROV PANEL (FIG.8)
G.B.T. DEPLOYMENT CONTROL SYSTEM.

# Control system facing up to new constraints

VIGNE
CA – France

Subsea developments in new fields must overcome the difficulties generated by deeper sites, greater distances and restricted budgets. Taking into account these constraints, conventional techniques and components quickly reach their limits. It is therefore of prime importance that the future be prepared by completely new concepts.

By replacing the usual electrohydraulic control pod by an electrohydraulic network linking standard light modules; the TOV system allows for a very reliable, ROV operable control system. The module, being dedicated to the basic function (control of one valve), becomes the standard and considerably decreases the amount of engineering needed by a Xmas tree equipment. Furthermore, the logistic suport is extremely limited since no heavy running tool is required and spare parts are kept at the module level.

The outstanding characteristics of the TOV :

- depth rating 2000 m.
- weight : 30 kg
- communications up to 1 Mb/s
- reliability (1 million hours MTBF), and therefore reusable

makes the module adpated to every condition.

As a complement to the TOV concept, ECA has developed components for special conditions (subsea mateable optical connector for long range transmission, subsea HPU to economise on hydraulic umbilicals ...).

## 1 FOREWORD

As soon as the engineering phase related to an offshore oil or gas field starts, the main question is "subsea or not subsea"? Usually considered as an expensive solution, not without reason in the past but ... things are improving - subsea is chosen only when it is really impossible to do otherwise. Tomorrow, it is almost sure that the subsea solution is to be the leading one. Consequently, the proposed components should be ready to cover the new needs and, in particular for control systems, the new technologies made available from other fields of activity should be used in accordance with the trends emerging today.

These trends can be summarized in three words : **cheaper, deeper, further** :

"Cheaper", or less expensive, depending on the position of the reader, is the master word illustrated by the CRINE initiative in the U.K.

"Deeper" is becoming more and more current and the new finds are situated down to several hundreds of metres mainly in Brazil (world leader 1027 m), Gulf of Mexico, and even in the U.K. in the West of Shetland area.

"Further" has been made possible by improvements in the transport of hydrocarbons, including boosting units, and control systems must be able to meet the new requirements.

The solution to these new trends are proposed by ECA in the form of :

- TOV system : lightweight, modularity and standardization answer the requirements "cheaper and deeper"

- Subsea HPU answers the requirement "cheaper"

- Subsea mateable optical connector answers the requirement "further".

## 2 TOV SYSTEM DESCRIPTION

### 2.1 Motivation

The TOV system (standing for TeleOperated Valve) is a new concept of a control system which main objectives are cost reduction and deep water access. Consequently, the effort is concentrated on the subsea part of the system.

Today, motivated by technical and economical aspects (cost of umbilicals to be decreased), the sophistication of the subsea systems led to size and weight increases. Furthemore, the concentration of underwater connectors on a single pod requires improved positioning accuracy. All these parameters have a bad effect on cost, ease of intervention, and reliability. A usual control pod, rated for 150 metres diverless use, is currently 2 cubic metres "fat", weighs a couple of tons, requires heavy installation tools (in turn remotely controlled) and has, when fully redundant, a 2 to 5 years MTBF. The TOV, presented hereafter, improves this situation.

### 2.2 General organization of the TOV system

As indicated before, emphasis has been put on the subsea part of the system. Consequently, only this part is described, the surface unit and the umbilicals remaining in the same configuration as for a usual control system. This means, as well, that any solution aiming at the suppression of umbilicals is usable with the TOV system.

The general organization relies on the fact that the TOV system is a distributed control system as opposed to the centralized pod. To ensure the control of a Xmas tree, one TOV module per actuator to be controlled is necessary. Each module is linked to the "electrohydraulic network" constituted by the umbilical. Figure 1 illustrates this arrangement for a single Xmas tree.

In order to remain compatible with any field organization, and to use a minimum number of umbilicals, dispatcher units have been developed, either active or passive. Electrical and hydraulic power are always passively distributed (active units on request). Data distribution depends upon the number of TOVs connected, length and characteristics of umbilicals.

## 2.3 Organization on a Xmas tree

Small size and weight of the TOV module allow its installation at any location in the Xmas tree. Nevertheless, in order to standardize the interfaces, a panel arrangement has been developed.

Hydraulic and electrical distributions between the umbilical and the TOVs (the network) are made within this panel and are not subject to subsea disconnection. The links between the panel and subsea valves and sensors are made by the integrator (usually the Xmas tree manufacturer) onshore.

By reducing to nothing the cable connections to be made subsea (except when the arrangement requires the umbilical to be connected to the panel via jumpers), the overall reliability is further increased.

## 2.4 Description of the TOV Module

### 2.4.1 General (see Figure 2)

The main design objectives have been the standardization (each module has the same hardware, thus allowing series manufacturing and cost reduction) and small volume and lightweight. For standardization, the TOV module functions have been defined as the same basic ones as that of a Xmas tree :

- control of one hydraulic valve,
- data acquisiton from two external sensors, an internal pressure sensor and eventually two limit switches indicating the valve position,
- network interface management,
- integrated safety functions and self-testing.

It is diver or ROV installable and retrievable, with adapted tools, by means of an electrohydraulic connector.

The module is pressure balanced, filled with dielectric oil and contains a small pressure resistant tank for the electronics.

### 2.4.2 Electronics

The electronic subassembly is fully contained in a pressure resistant tank (600 m standard, option 2000 m) which dimensions are 90 mm o.d, 180 mm length. It includes regulated low voltage power supplies from the network power and only one hybrid circuit (64 mm x 64 mm) ensuring all the functions.

This hybrid circuit has been specially designed for the TOV and manages :

- Communications with the surface unit via the network, using HDLC protocol, at a programmable speed from 1200 b/s to 1 Mb/s
- Opening and closing control of the TOV bistable solenoid valve
- Sensors, data acquistions, digitalization and calibration
- Integrated self-testing
- Local safety automatisms, which parameters are programmable via the network (valve closing in case of communication loss, sequential shut down ...).

The TOV electronics is independent from the application, only the software is dedicated. Some parameters are changeable during the operations, via the network.

The MTBF of this subassembly is $2.10^6$ hours.

### 2.4.3 Hydraulics

The hydraulic subassembly is mainly composed of one bistable solenoid valve, with two stages, the power stage being of a shear seal technology. Pilot stages are redundant and pulse operated.

A safety device ensures the valve closes in case of loss of hydraulic pressure. The output to the Xmas tree valve is filtered (with additional pressure relief valve) to avoid solenoid valve pollution by the Xmas tree actuator.
This output is equipped with a pressure sensor allowing the monitoring of both the pressure and valve position by pressure footprinting.

Water based or mineral fluids are usable with a maximum service pressure of 5000 psi (optionally up to 11000 psi, 15000 psi in progress).

The solenoid valve is sub plate-mounted on a drilled block, including filter and pressure relief valve and self-sealing couplers (hydraulic input, valve output and, eventually return port).

### 2.4.4 Physical characteristics and performances

The TOV module features the following characteristics :

- Size : length 290 mm (420 mm with locking device)
  width 130 mm
  height 280 mm (310 mm with couplers)

- Weight : 35 kg in air
  30 kg in water

- Connection : by diver or ROV with dedicated tool

- Depth : standard 600 m
  option 2000 m (same size and weight)

- Hydraulic pressure 5000 psi
  option up to 11000 psi (15000 psi in progress)

- Hydraulic fluid : water based or mineral

- Cleanliness class of the network : 8 (NAS 1638)

- Electrical consumption : permanent 4.3 VA
  max 30 VA

- Communication speed : 1200 b/s to 1 Mb/s (standard 64 kb/s)

- Maximum distance without modem to be computed depending on the configuration
  example : 1 well 32 TOVs : 10 km with 2.5 mm cable

- Data acquisition : accuracy 0.1% FSR
  period : 40 ms
  option : faster acquisition with decreased accuracy.

2.4.5 Installation and retrieval

One of the main advantages of the TOV module is its physical characteristics which allow easy installation and retrieval by ROV, or even by diver in shallow waters, using a very simple and light (50 kg) interface. Guide lines are no longer necessary, nor the heavy running tool.
In addition, its design for 600 metres water depth, optionally 2000 metres, makes this concept the ideal solution for deep water control.

## 2.5 Advantages

Besides the advantages given by the technical characteristics (lightweight, operational in deep water, reliability ...), the TOV concept presents additional advantages directly contributing to cost reductions, both in capex and opex :

- The standardization of the modules, the interfaces and the functions lead to a very light engineering related to the application, compared to the engineering for a traditional control system.

- The basic function dealt with by the module allows the best adaptation of the system to the application. (A TOV system for 4 valves is twice cheaper than the one for 8 valves).

- The spare part for a complete system remains at the level of one TOV only and the tools are in the same range.

## 2.6 Status of development

The TOV system exists. Several units have been built and will be tested underwater in Norway from April to October 1996 with the help of the EEC, under a Thermie programme. Several oil companies are associated with this project.

Prior to the launch of this long lead demonstration, inhouse tests have taken place in order to check the main functionalities within the specified environment. In particular, mating and unmating tests have been carried out underwater by divers using a real panel and a dedicated installation tool.

## 3 SUBSEA HPU

As previously stated, subsea HPU is a "cheaper" solution and can be used together with the TOV system. Cheaper, because the umbilicals in a control system represent a wide part of the capex. They are usually composed of :

- two redundant high pressure (HP) lines
- two redundant vey high pressure (VHP) lines.

The supply alone of these lines represents several £ 10000 per kilometre, figure to be doubled taking into account the installation.

By suppressing these lines a million pounds is quickly available. A complete subsea HPU, being far less expensive, therefore presents a real interest. Taking advantage of the TOV subsea demonstration, such a subsea HPU has been built and will be operational on the same test bench. This HPU will be composed of :

- a subsea reservoir (500 l.)
- an accumulator bank
- a subsea pump.

This last unit is described in more detail hereafter.

The unit is composed of a pump and an electrical motor installed in a pressure balanced vessel (300 mm o.d., length 700 mm.). It is rated for 600 metres depth. Both mineral and water based fluids are usable. An electrical converter supplies the motor with the necessary voltage to reach the programmed output pressure.
Two levels of output pressure are programmable, allowing the generation of HP and VHP within the Xmas tree. The maximum pressure is 10000 psi.
From an external control the output pressure is switched, together with an external valve, allowing the right accumulator bank to be fed.
During the demonstration, this unit will be used with Marston Bently HW 540 fluid.

## 4 SUBSEA MATEABLE OPTICAL CONNECTOR

Here is the "further" solution. Transmitting data over several tens of thousands of metres is not as simple as merely stating it. Moreover, in the case where the power transmission is dedicated to a subsea pump, it is virtually impossible to use any electrical support. For reliability of transmission therefore the solution is : optical fibre.

Commonly used onshore and underwater, the optical fibre has been made the standard transmission medium by the telecommunications industry. This has resulted in a fall in prices down to the level of the copper pair price.

The fibre optic transmission is well mastered technically and offers an exceptional reliability over very long distances without any amplifcation. For an oil and gas application, the control unit being retrievable, an underwater mateable connector is necessary.

The underwater mateable optical connector is based on an existing dry mateable, underwater optical connector designed and manufactured by ECA for its minehunting ROV (several thousands of connectors operational). The technical basis is that of the expanded beam, as each extremity of the fibre is equipped with aspheric lenses providing expanded collimated beam, thus ensuring a very good immunity to pollution. Around this optical heart a suitable mechanical housing has been built to make the connector easily mateable either by diver or by ROV. The main features are :

- Aspherical lenses protected by glass windows,
- Pressure-balanced design with optical liquid
- Subsea mateable down to 600 metres
- Mating and unmating by gravity, ROV or diver
- Permanent sealing when disconnected.

The optical performances are an insertion loss of 1.5 dB at 1300 nm and 1550 nm with single mode fibre.

## 5 CONCLUSION

This presentation shows that the new trends illustrated by the seemingly non-compatible three words "cheaper", "deeper", "further" can lead to the use of advanced technologies with a view to finding sophisticated and effective solutions for the control system. These solutions already exist and are about to be qualified by a long-term underwater test :

- The TOV distributed control system
- The subsea HPU using water based fluid
- The underwater mateable optical connector.

**Figure 1 - General organization for a Xmas Tree**

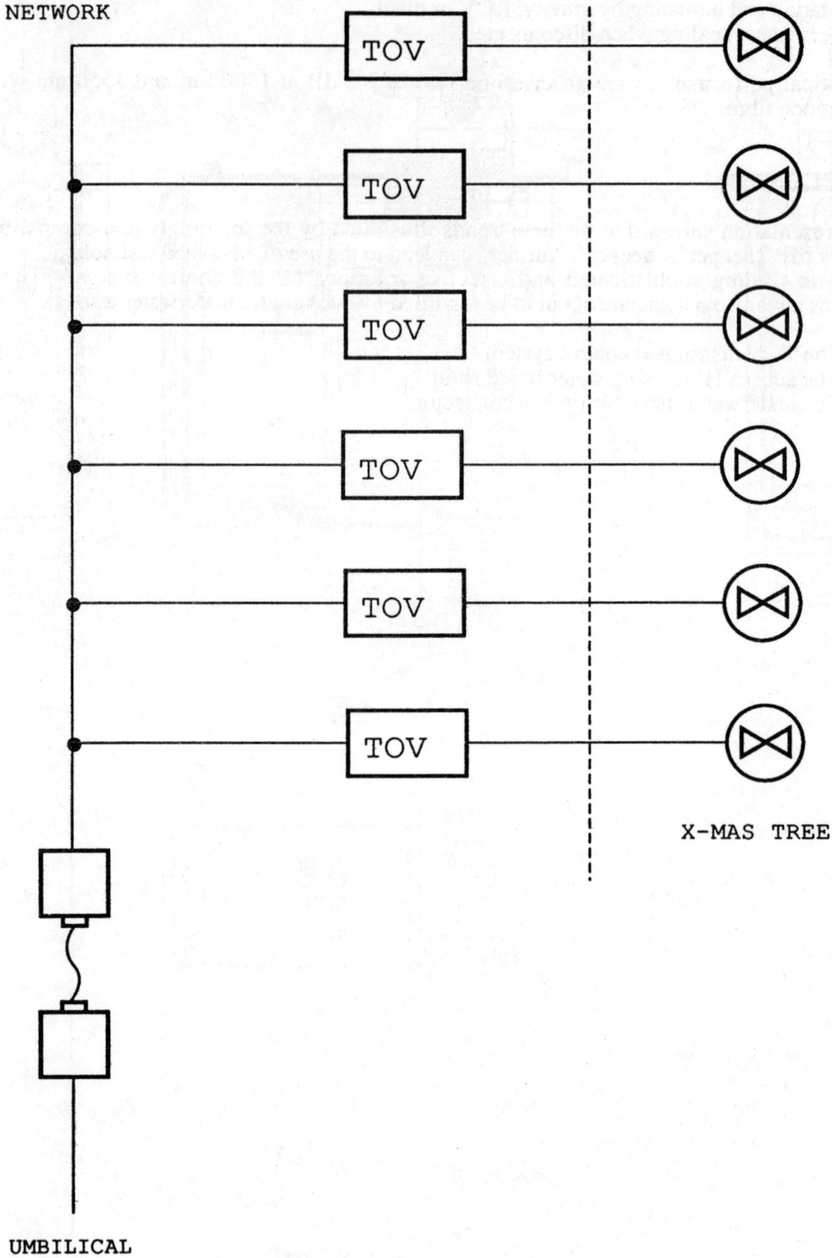

## Figure 2 - TOV Module

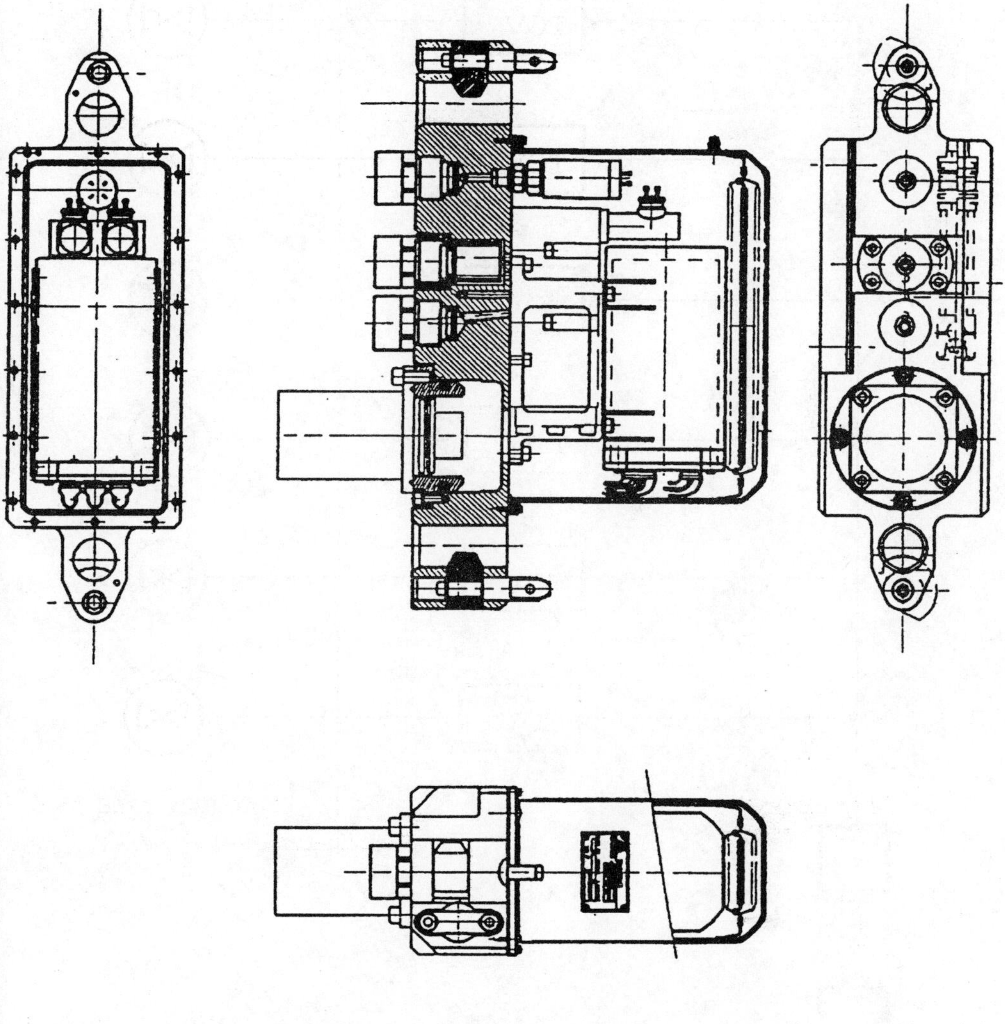

## Service Experience

# ubsea hydraulic pressure intensification – into ervice

ENJAMIN
B Seatec

**ABSTRACT**

This paper provides an update on the development of a high reliability subsea High Pressure Hydraulic Intensifier (HPI), and its successful introduction into field service.

A brief history of the device begins with the origins of development, and continues by recounting some details of the Technology Development Programme which formed Phase 1 of the project.

Discussion of Phase 2 of the programme is segmented into Phase 2a (Prototype Design), Phase 2b (Development and Test) and Phase 2c (Productionisation). The issue of achieving the required reliability by means of a combination of both mathematically-based analysis and bench testing is considered.

The paper concludes with a description of the first field service application, a review of the lessons learned from the programme, and a forward look at other possible applications.

Both the Technology Development programme and the initial design work for the HPI were carried out for GEC Marconi Oil & Gas by GEC Alsthom's Mechanical Engineering Centre, and the programme was supported by two grants from the Department of Trade and Industry's Offshore Oil and Gas Supplies Office (OSO).

Prepared by: Roy Benjamin
Chief Mechanical Systems Engineer
GEC-Marconi Oil & Gas
2 High Street
Nailsea
Bristol
BS19 1BS

Telephone: (01275) 810100
Fax: (01275) 851467

## 1. INTRODUCTION

In 1984 the U.K. Oil & Gas Projects and Supplies Office (OSO) commissioned GEC Marconi Oil and Gas (G-MOG), to undertake a study into umbilical-free subsea production control systems. This study showed the feasibility of developing a fully autonomous satellite wellhead control package utilising through-water acoustic telemetry, combined with subsea electric power generation. In 1987 a further phase of this work, jointly funded by a number of oil companies, developed many of the core technologies which would be essential to the success of such a system. A significant part of the experimental programme centred around the design of a reliable subsea hydraulic power unit, during which GEC-Alsthom's Mechanical Engineering Centre (MEC) worked in close association with G-MOG.

In order to overcome the difficulty of sourcing a suitable high pressure zero-maintenance pump to serve the surface controlled sub-surface safety valve (SCSSV), a scheme was put forward whereby a reciprocating high pressure intensifier would be driven from the low pressure (3000 lbf/in$^2$; 207 bar) tree valve supply. Although originally conceived as part of an overall subsea electrical and hydraulic power generation scheme, subsea intensification has recently attracted renewed interest for the development of high pressure oil/gas reservoirs where existing umbilical technology falls short.

There is now a growing interest in the development of high pressure fields which presently require SCSSV control pressures of up to 760 bar (11,000 lbf/in$^2$). The control umbilical hose requirements for this pressure can just be met with existing hose technology without compromising the 4:1 factor of safety on burst pressure which is felt by the industry to be prudent. For some projected installations however, pressures of up to 1000 bar (15,000 lbf/in$^2$) will be required. A significant reduction in the factor of safety will thus be necessary if existing hose technology is to be used; this is something most operators are reluctant to contemplate due to the severity of the application, the areas of doubt which remain over issues such as fluid/hose liner compatibility, and above all the cost implications of hose failure and consequent lost production

One way of overcoming the difficulties associated with high pressure hoses is to eliminate them from the system altogether. This can be achieved by deriving the high pressure requirement at the sea bed by intensification of the low pressure tree valve actuator supply. The SCSSV control lends itself to this approach since the hydraulic energy requirements of the valve are minimal when compared with those of tree valve actuators. A typical SCSSV may require only 0.1 litres or so of fluid to operate the piston.

In addition to providing a technical solution to the demands of new high pressure fields, the use of an intensification scheme can also show very significant cost savings in terms of reduced topsides equipment and a cheaper umbilical. These cost savings are applicable to most satellite subsea production control systems, whether or not they are in a high pressure field.

In January 1992, MEC commenced work on the first phase of a two-phase programme to develop a high pressure intensifier suitable for use in a subsea wellhead control system. In the Phase I programme (entitled "Technology Development") the technology required to build a reliable device which could be installed in a subsea control module was developed with the aid of several test rigs on which the key component designs were endurance tested. This was followed by a larger Phase II programme in which prototype development led on to qualification of a production-standard subsea High Pressure Intensifier (HPI). Figure 1 shows the first prototype unit (P1).

From the outset, heavy emphasis was placed on the development of a highly reliable device. The central design target was 1 million cycles of operation without any significant degradation in performance.

## 2. PHASE I - TECHNOLOGY DEVELOPMENT

### 2.1. Technical Requirement Survey

In order to help gauge the market for a subsea high pressure intensifier, and to assist in the specification of a device which would have the broadest possible appeal, it was decided that operators' views should be canvassed by questionnaire. During early 1992, a questionnaire was sent to around 20 operators.

#### 2.1.1 Pressures

From the questionnaires returned, and the conversations which ensued, it was clear that a 15,000 lbf/in$^2$ (1035 bar) capability would meet all new field development requirements for the foreseeable future.

#### 2.1.2 Flowrate

As previously stated, the flow required to open an SCSSV is minimal. The governing factor is leakage downstream of the intensifier. Where respondents were able to indicate the typical leakage which might be tolerated from the SCSSV before intervention, the figures given showed considerable variation. However, assuming that zero leakage is unlikely to be achieved in practice, and assuming that very high leakage would be logistically unsustainable by most operators, a maximum leakage rate of 3 l/day over the life of the intensifier was considered to be a realistic design value. (It should be noted that the intensifier which is described in this report is actually capable of producing flow rates of 0.5 l/min or more, limited only by the ability of the low pressure system to supply flow via the umbilical. Some reduction in unit life might however be expected under such circumstances.)

#### 2.1.3 Fluid

All of the respondents indicated that the intensifier would need to be capable of operation on water based fluids, with a cleanliness level of typically NAS 1638 Class 6/8, but the ability to operate on class 10 or 12 fluid.

### 2.2 Device Configuration, Specification and Control

#### 2.2.1 Configuration

The most common type of hydraulic intensifier employs a differential area piston, and this was judged to be the most appropriate configuration in this instance. A double-acting device which pumped in both directions of travel was chosen, to minimise possible pulsation effects.

After considering many possible schemes for intensifier reciprocation, an arrangement was selected which employs a hydraulically piloted changeover valve in order to give maximum life with minimum mechanical wear. Furthermore, it was decided that the changeover valve together with its pilot valve should employ shear seal principles, partly because this is the accepted industry standard for subsea hydraulic valves, and partly because such valves exhibit very low levels of leakage which is an important consideration for intensifier efficiency.

### 2.2.2 Functional Design Specification

A detailed Functional Design Specification was developed, the leading particulars of which were:

- Modular construction (comprised of intensifier; changeover valve, optional dump valve).
- Shear seal design.
- Compatible with both upstream and downstream Directional Control Valves.
- Design adaptable to different intensification ratios.
- Design Life – 15 years for mainstream applications
  – 25-30 years for special installation
  – 1 million cycles without significant wear
- Target Service Interval – 5 years minimum
- Nominal Supply Pressure – 207 bar (3000 lbf/in$^2$)
- Maximum Delivery Pressure – 1035bar (15 000 lbf/in$^2$)
- Ambient Seawater depth – 1000 metres (3281 ft)
- Control Fluid – Synthetic mineral oil or water-glycol
- Control Fluid Temperature – –20 to + 50°C
- Air Temperature (storage) – –20 to + 50°C
- Cleanliness of Fluid – Designed for operation with fluid not less clean than NAS 1638 Class 10

### 2.3 SCSSV Control Options

The HPI may be sited either upstream or downstream of the Directional Control Valve(s) - (DCVs). If the DCV is upstream of the HPI, it only sees low pressure; also, the HPI duty is less arduous, as it is only at pressure when the DCV is open. Conversely, if the HPI is upstream of the DCV(s), a single intensifier may be used to supply several SCSSV's. Either way, the HPI must be used in conjunction with a dump valve to allow pressure to be vented in an emergency.

### 2.4 Design Details

#### 2.4.1 Schematic Arrangement

Fig 2 shows the schematic hydraulic circuit for the intensifier. The device comprises a pair of opposed dual area piston assemblies (1) and (2), between which is a pilot valve section (3). The pistons bear directly onto the slider of the pilot valve but are not attached to it. Hydraulic fluid is drawn into one or other of the high pressure piston chambers (area A1) from the low pressure supply line (LP) via the inlet non-return valves (4) or (5). After intensification the fluid is expelled from one or other of the high pressure piston chambers via the outlet non-return valves (6) or (7). The pistons and pilot valve slider are reciprocated as an assembly under the action of a change-over valve (8).

### 2.4.2 Practical Arrangement

Fig 1 shows the mechanical configuration of the intensifier in which the change-over valve is mounted on top of the central pilot valve section. Both change-over and pilot valves are of the "shear seal" type, with precision ground seal elements sliding on a precision ground flat-faced slider bar. In conventional shear seal valves of 4-way configuration, 3 seal elements are often used in line along the axis of the slider to effect the change-over porting. The stroke of the pilot valve slider is governed by the working stroke of the intensifier, and for a usable swept volume the conventional 3-in-line seal element arrangement results in an unacceptably long device. To circumvent this problem, a novel design configuration has been adopted, comprised of a square section slider bar in which shear seal elements act on all four faces. This allows the axial length to be substantially reduced. Fig 3 gives an indication of overall envelope size for installation purposes.

### 2.4.3 Patent

The device described above is the subject of a UK patent application number 9304092.1.

## 2.5 Mathematical Modelling of Intensifier

In order to determine how the proposed intensifier design might behave when operating an SCSSV, a mathematical model of the device was constructed using ACSL (Advanced Continuous Simulation Language). Fig 4 shows the schematic circuit for the model, in which the load comprises an R-C-L network representing a typical 0.25 inch OD SCSSV control line, plus parallel capacitances to represent the trapped volume in the actuator and the spring compliance term. The changeover valve is switched from the intensifier piston position, with suitable time constants introduced to simulate the delays resulting from the pilot line pressurisation and de-pressurisation.

Fig 5 shows the outlet pressure from the intensifier building up to 1000 bar (15 000 lbf/in$^2$) from an inlet pressure of 207 bar (3000 lbf/in$^2$), with 8 – 9 pumping cycles required to pressurise the control line and displace the SCSSV piston. The control line forms a very effective filter for the ripple on the intensifier outlet pressure, as can be seen from Fig 6.

Although a number of approximations were necessary to enable a simplified model to be constructed during Phase I of the project, the simulation work proved particularly useful in verifying the chosen piston areas and stroke to give low velocities and hence increased cyclic life from the high pressure seals. A more comprehensive computer model was developed under the Phase II programme in order to prove the switching and latching arrangement prior to cutting metal.

## 2.6 High Pressure Seal Development

In order to investigate and compare the performance of a number of candidate high pressure seals, a test fixture was manufactured. This fixture comprised a three-section cylinder body designed to accept a stepped piston having a 5:1 area ratio. The test seals were all sufficiently close in size to enable the same seal housing section to be used for each, and it was necessary only to change the bronze anti-extrusion back-up ring to accommodate slight variations in axial length.

Three different candidate HP seals were tested (with two different piston rod treatments) before arriving at a combination which completed 150 000 cycles without showing any signs of seal wear.

## 2.7 Shear Seal Development

A shear seal test fixture was built to cycle a pair of opposed shear seal pistons. The complete shear seal fixture was mounted onto a stainless steel base, and submerged in a tank of water-glycol hydraulic fluid.

A number of shear seal designs was evaluated before arriving at a well-balanced design which met all of the relevant design criteria. The final design was cycled for 100 000 cycles, checking for seal leakage every 25 000 cycles. On each occasion the leakage was immeasurable over a 5 minute period with the slider bar static and an applied pressure of 207 bar. After completing the test, the surface finish of the slider bar was measured, and found to be unchanged. The test was then continued for a further 50 000 cycles, after which the leakage was again found to be immeasurable. As a final test of the design the fixture was operated on town mains water for 25 000 cycles, again with satisfactory results.

## 2.8 Check Valve Development

Many different designs of check valve were contemplated before deciding that the most appropriate type of sealing element would be a ball on a chamfered seat. This, it was felt, would provide the most reliable and leak-free solution with little chance of malfunction, and a fast response to improve intensifier volumetric efficiency.

A large number of tests was carried out using a variety of different materials, finishes and seating configurations before a satisfactory combination was found which gave 100 000 cycles without failure.

## 2.9 Phase 1 Review

The technical requirement survey which was carried out at the beginning of Phase 1 clearly showed the trend towards higher SCSSV control pressures, with 1035 bar (15 000 lbf/in$^2$) being a typical future requirement for most of the operators who responded to the survey.

The intensifier design evolved in Phase I takes advantage of existing subsea technology; it is now the generally accepted wisdom that shear seal valves offer the most suitable combination of properties for subsea use, such as very low leakage, high contamination tolerance, automatic wear compensation and long life. By using shear seal-type pilot and changeover valves together with hydraulic latching, a device can be built which will offer high reliability and long cyclic endurance, both characteristics which are essential in a subsea, minimal intervention situation.

The mathematical modelling which was carried out suggested that the impedance of the control line to an SCSSV would act as an effective filter to the pulsations caused by piston reciprocation.

The three aspects of the proposed intensifier design which were considered to represent the greatest technical risk were investigated in the laboratory on a series of specially built component test rigs.

All three test programmes proved technically demanding, but after some iterative redesign and changes of material, the design target of 100 000 cycles for each of the three rigs was comfortably achieved. This work paved the way for the Phase II development and qualification programme.

## 3.0 PROTOTYPE DEVELOPMENT AND PRODUCTIONISATION

### 3.1 The Programme

Initial proposals for a single-phase JIP programme embracing prototype development and qualification testing attracted considerable interest from oil companies but insufficient funding was secured to run the project as planned. To preserve momentum it was therefore decided to segment the work into a Phase 2A (Design) and a Phase 2B (Prototype Development and Test), with a possible Phase 2C (Productionisation), the latter being dependant upon a specific field application emerging.

### 3.2 Phase 2A - Prototype Design

Phase 2A consisted of three essential elements:-

- More detailed mathematical modelling to:-
  - verify consistent intensifier operation under all conditions.
  - obtain data for pilot and changeover valve port spacing and laps.
  - investigate the effect of trapped air on operation.
  - verify sizes of changeover valve spring, pistons, and need for check valves.
  - verify sizes of drilled galleries within unit.
- Detailed design scheme followed by piece part manufacturing drawings.
- Parts manufacture and assembly.

This six-month programme of work commenced in early 1994, supported by a further grant from the OSO. The emphasis on risk reduction inherent in Phase 1 began to show early dividends with no major problems encountered during simulation or detailed drafting. Considerable emphasis was placed on minimising the size and weight of the unit and on ensuring a highly reliable design to achieve the one million cycle target (see Section 2.2.2).

By the Autumn of 1994 design was complete and all parts were on order. All work to this date had been aimed at a 5:1 pressure transformation ratio unit, converting a 207 bar (3000 lbf/in$^2$) supply into a 1035 bar (15 000 lbf/in$^2$) output.

### 3.3 Phase 2B - Development and Test

Up until now, work had been conducted on a low-key development basis. However, towards the end of Phase 2A the situation changed because a field development application emerged in the form of Saga Petroleum's Vigdis Project. The client's own studies had identified the benefits of a subsea HPI for installation in the control pod - but with a pressure transformation ratio of 2.5:1. Clearly this was too good an opportunity to miss, so a rapid re-appraisal of the development project was conducted to establish:-

a) The implications of reducing the output pressure by half.

b) The risk, timescale and cost factors implicit in an accelerated, project-orientated development programme requiring production-standard HPI units within a relatively short timescale.

Again, the benefits of the Phase 1 Technical Demonstration Programme manifested themselves, partly because the modular design approach allowed the different intensification ratio to be easily accommodated and partly because confidence was high that the critical design elements (high pressure shaft seals; shear seals; check valves) had already been proven.

The decision was therefore taken to re-focus the project on the Vigdis application and to carry out development, test and productionisation at GEC-Marconi Oil & Gas's facility at Nailsea, Bristol. Phase 2B now involved:-

• Modifying some parts to de-rate the output pressure for a 517 bar (7500 lbf/in$^2$) first prototype (P1).

• Ordering further sets for a second 517 bar prototype unit (P2) for cold tests and stall tests.

• Designing and building a test rig to accept both P1 and P2 units.

It was also decided to retain the original design objectives by procuring long lead items for a P3 unit which would retain the 5:1 pressure transformation ratio, giving an output pressure of 1035 bar (15 000 lbf/in$^2$). The test rig design parameters were maintained at this high pressure.

Assembly of P1 was completed on schedule in early 1995. It was very gratifying for the engineering team to find that this unit operated exactly as intended the very first time pressure was applied. Some teething problems subsequently emerged which proved difficult to trace but relatively easy to solve. In summary these were:-

| PROBLEM | SOLUTION |
|---|---|
| 1. Hydraulic lock in changeover valve. | Radial pressure relief groove added. |
| 2. Non-uniform shaft seal wear in changeover valve. | Additional bearing surface added. |
| 3. Unacceptable rate of piston seal wear in mainstage valve piston. | Chromed bores. |
| 4. As above. | Harder seal material. |

The first two modifications listed above resulted in a run-life of around one-quarter of a million cycles. Modification number 3 extended this to approximately three-quarters of a million cycles for the mainstage (the changeover valve having already achieved the design target of one million cycles). Modification number 4 is being implemented at the time of writing (November 1995) and we are confident this will allow us to comfortably exceed our design target.

Also as this is being written, P1 has been reassembled with the new seals (Modification no. 4), and P2 is ready for cold testing and stall testing. Both units will be run for at least one million cycles. Most P3 parts (1035 bar/15 000 lbf/in$^2$) are available.

## 3.4 Phase 2C Productionisation

Whilst our internally set design life target was one million cycles without significant wear, a figure of one-quarter of a million cycles was considered adequate for the Vigdis application and, having reviewed the results from the prototype test programme, the Vigdis project team accepted the design for production in the late Spring of 1995. By October 1995, the twelve production-standard HPIs for Vigdis and the associated production test rig were in final assembly.

## 4.0 FIRST FIELD SERVICE APPLICATION

Figure 7 shows the hydraulic schematic for a typical HPI control pod installation in which the HPI and associated accumulator and dump valve may be identified. Note that if required, it is possible to incorporate a cross-connection arrangement between a pair of pods to provide redundancy.

Deliveries of Vigdis pods will commence towards the end of 1995 and the first pod will be installed subsea in the Spring of 1996. A second field service application has now been secured in the form of Repsol's Poseidon field. This has an unusually long offset distance of 45km so the cost benefit of eliminating the high pressure hose from the umbilical becomes very significant, especially since well pressure is expected to decline rapidly, making a separate high-pressure supply unnecessary after only a few months production. This particular variant of HPI only requires a modest pressure transformation ratio of 1.33, giving an output pressure of 276 bar (4000 $lbf/in^2$).

## 5.0 ACHIEVING A HIGH RELIABILITY

## 5.1 Design & Simulation

Because the subsea HPI is intended to be installed in remote inaccessible locations, high reliability is of paramount importance. Stringent targets were set from the outset as defined in Section 2.2.2. Some explanation of these targets is perhaps necessary:-

a) Because subsea oil fields typically have lives of 10 - 15 years (and sometimes as long as 25 - 30 years), the hardware has to be capable of lasting this long, even if it is not used very often - see below.

b) The primary use of the HPI is to supply fluid to the SCSSV. Even if the HPI itself is designed for high efficiency and low internal leakage, this can be completely negated by a significant external leak rate. This demands the ability to operate for a high number of cycles with a very low wear rate and explains the high target figure of one million cycles.

c) If there is negligible external leakage and/or the SCSSV is only operated very occasionally, the HPI will either be stalled, or will creep very slowly under pressure. The ability to move when required after a long period of inactivity, together with stick-slip problems at low creep rates, each present their own set of design problems.

All these factors were taken into account right from the start of the design process. Extensive use of computer simulation was made to verify operation in all known conditions and the three Phase 1 Test Programmes were targetted at these critical design areas to ensure robust and reliable performance. Material selection, surface finishes and hard coatings all received very close attention and it is worth recording that the achievement of an acceptable level of performance proved an unexpectedly stiff design challenge during this first phase of the project.

Extensive prototype testing has been and is being used to steadily extend the design life up to and beyond the one million cycle target, with two prototypes being employed to minimise the chance of a 1-off test result.

### 5.2 Numerical Analysis

In addition to all of the above, numerically-based reliability analysis techniques were employed to identify an MTBF (Mean Time Between Failures) figure for the HPI which could be used in systems modelling for the complete subsea production control suite.

Long experience of reliability modelling for subsea production control systems has led us (and others in the oil and gas industry) to conclude that:-

a) theoretically-derived MTBF values for components are invariably much lower than those obtained from field service histories.

b) the principal benefit of the numerical approach lies in **comparing** the merits of alternative system designs, rather than deriving an absolute value.

In the case of the HPI, there is no directly comparable field service history on which to draw, but the well-known and widely-used subsea Directional Control Valve (DCV) uses analogous shear seal technology and may be used as a point of reference. Applying a similar approach to that used for DCV's gives an MTBF value for the HPI of around one third of the value for the DCV which is felt to be intuitively correct, given the relative complexity of the two components.

## 6.0 LESSONS LEARNED

### 6.1 Spending Money to Save Money

It often happens that a new product development goes seriously awry either technically or financially or both. In the case of the HPI development we have managed to avoid significant problems in either area and we believe there is no magic in this.

It is simply that, supported by the OSO, we made the decision to spend a significant sum of money on a set of Phase 1 technology demonstration (i.e. risk reduction) activities **before** we set a fixed budget and timescale for the main development programme. There is absolutely no doubt that this paid handsome dividends in Phase 2. The case for spending money to save money is often a difficult one to argue but every effort should be made to do so if our experience is anything to go by.

### 6.2 Broadening the Business Case

Nowadays, research and development decisions must be as fully justified by a return on investment financial analysis as they are by the technical innovation content. For specialised, low volume business such as Subsea Production Control Systems this can severely inhibit if viewed too narrowly.

Investment in HPI would be difficult to justify solely in terms of the market size and value of these as components. However, viewed as an enabling technology HPI allows systems solutions to be offered which are competitively superior, or where there is no other viable solution. Correctly basing the investment case on the additional systems business which the HPI technology enables generates a significantly different and viable cost justification.

## 7.0 OTHER APPLICATIONS

Interest has been shown in HPI by other branches of industry and naturally we welcome this. However, the design of HPI is aimed at subsea applications, and this is the most likely area of use.

There is considerable interest in High Pressure/High Temperature (HP/HT) reservoirs and HPI is aligned to this trend already, with a design working pressure of 15 000 $lbf/in^2$ (1035 bar). Some applications may require even higher pressures than this, and we see no major impediment to increasing the output rating to 20 000 $lbf/in^2$ (1380 bar) if required.

Another possible application for HPI is to pump chemicals. Because of the hostile nature of some of the chemicals used subsea this is not an issue to be treated lightly - but there would be obvious benefits in remote or HP/HT fields.

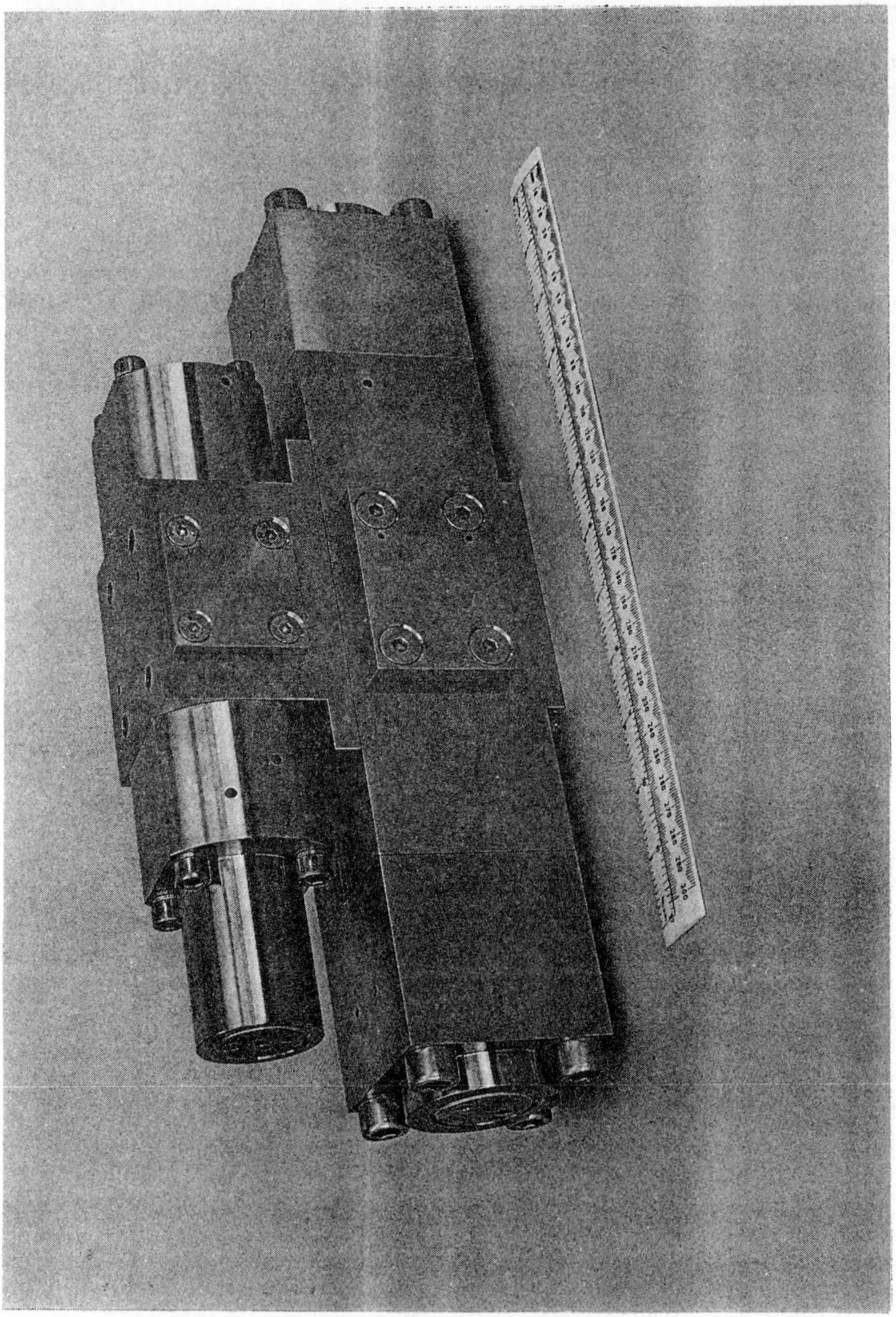

FIGURE 1 — PROTOTYPE HIGH PRESSURE INTENSIFIER

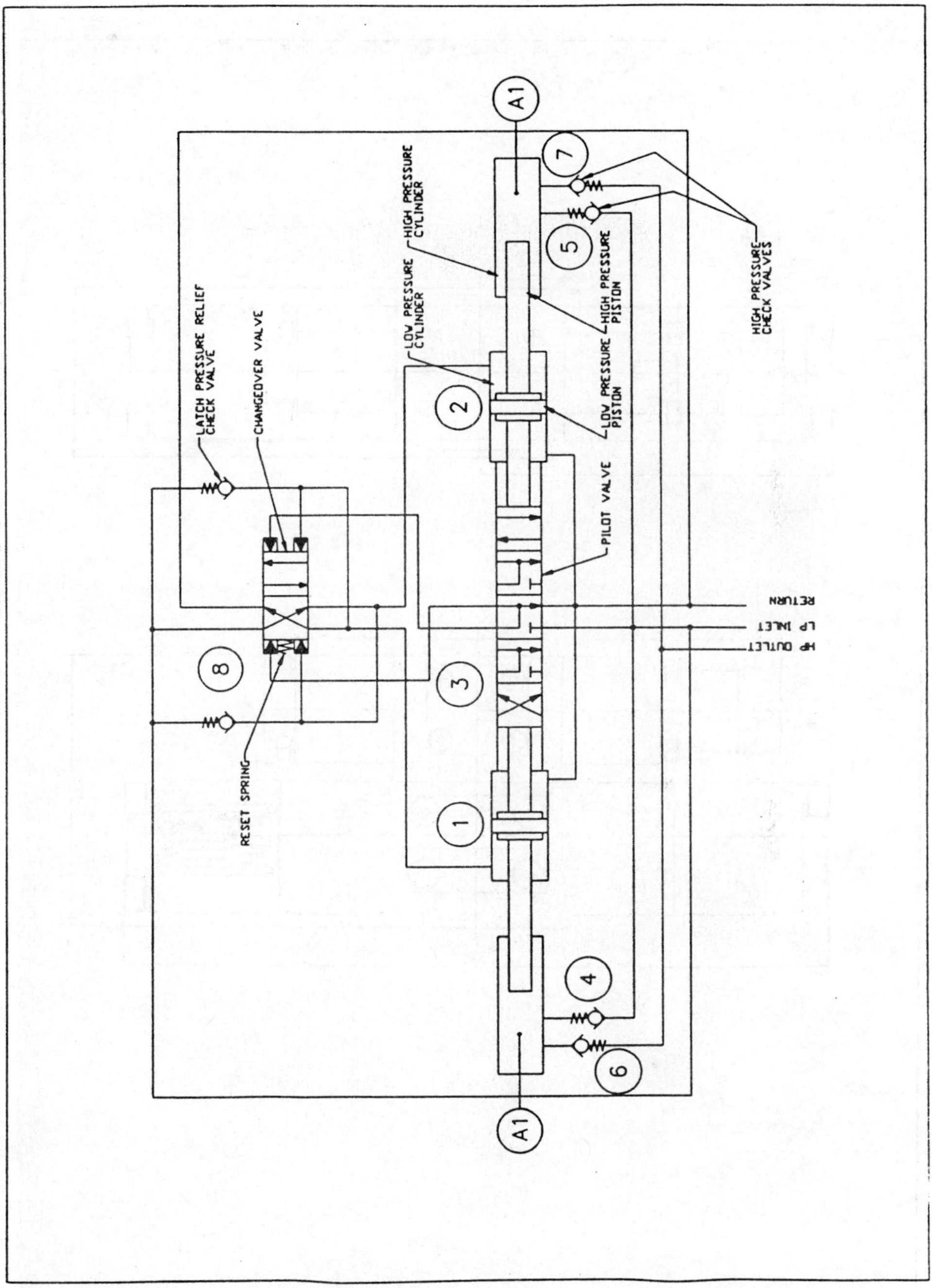


**FIGURE 2 — HP INTENSIFIER SCHEMATIC CIRCUIT**

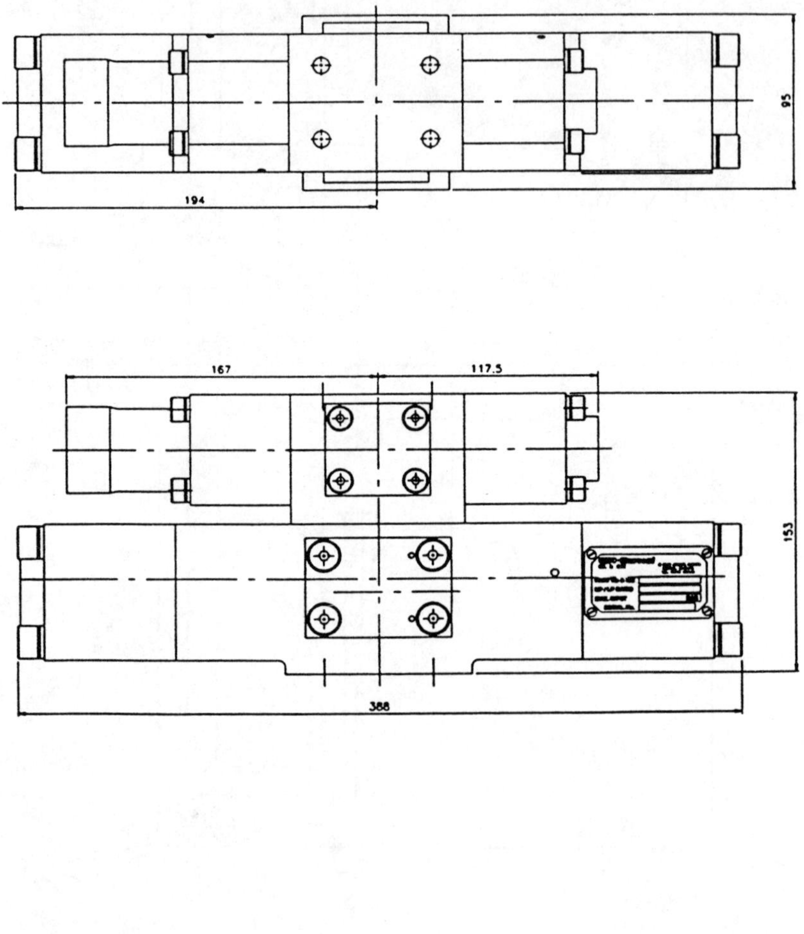


**FIGURE 3 — HPI SIZE ENVELOPE**

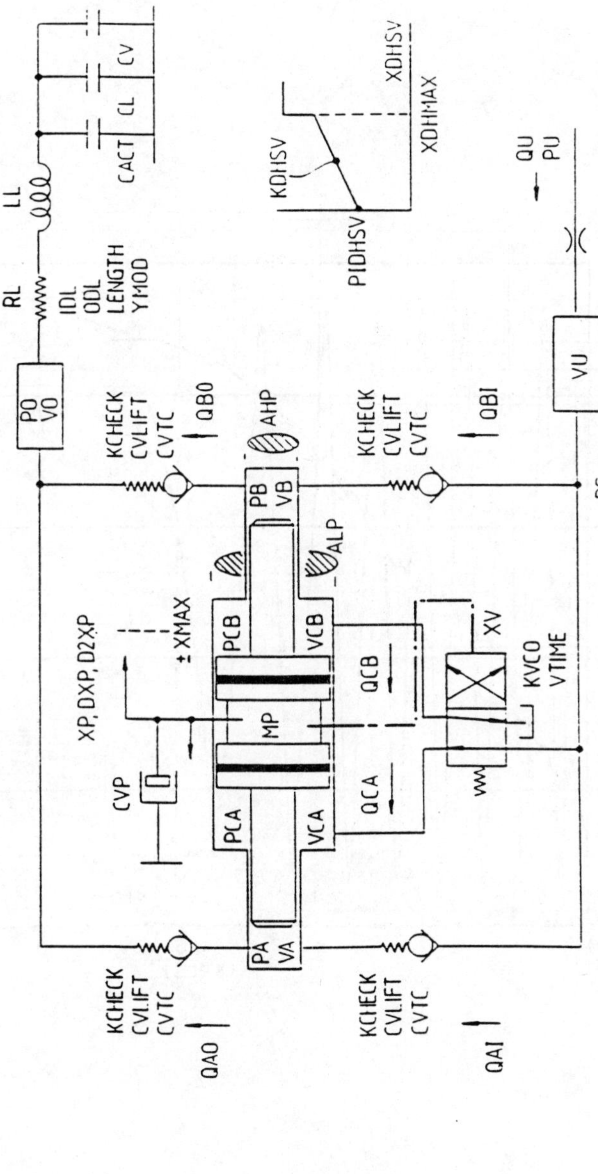


**FIGURE 4 — HP INTENSIFIER SIMULATION SCHEMATIC DIAGRAM**

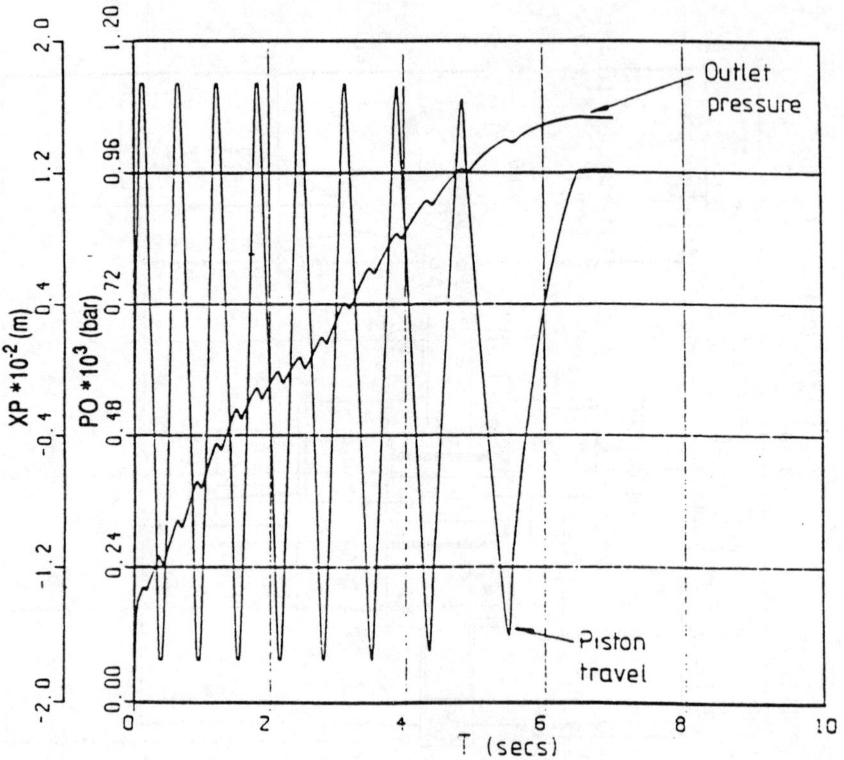


**FIGURE 5— HP INTENSIFIER - SIMULATED PERFORMANCE**

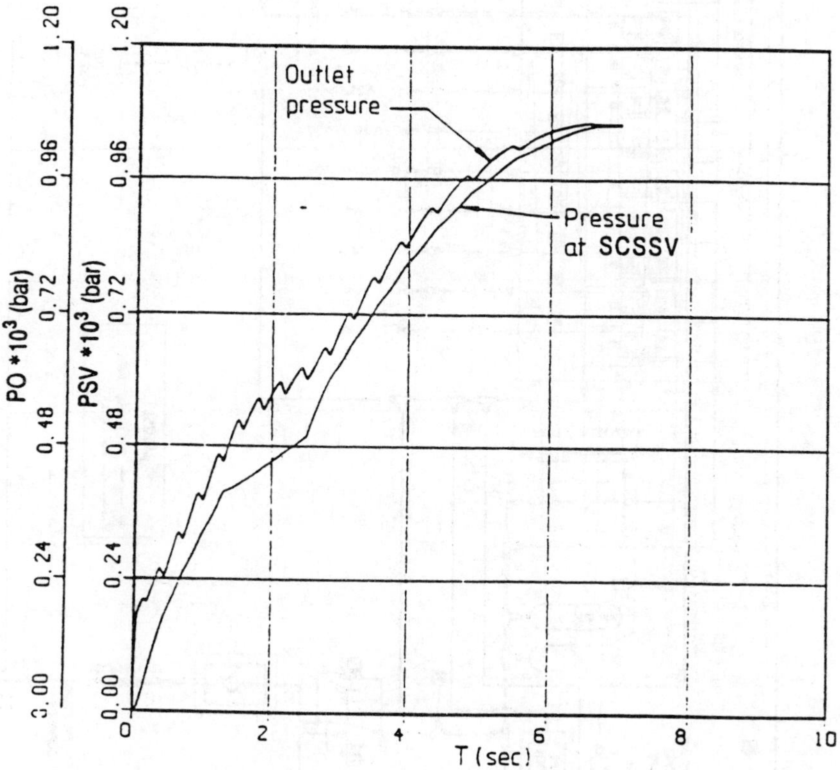


**FIGURE 6 — HP INTENSIFIER - SIMULATED PERFORMANCE**

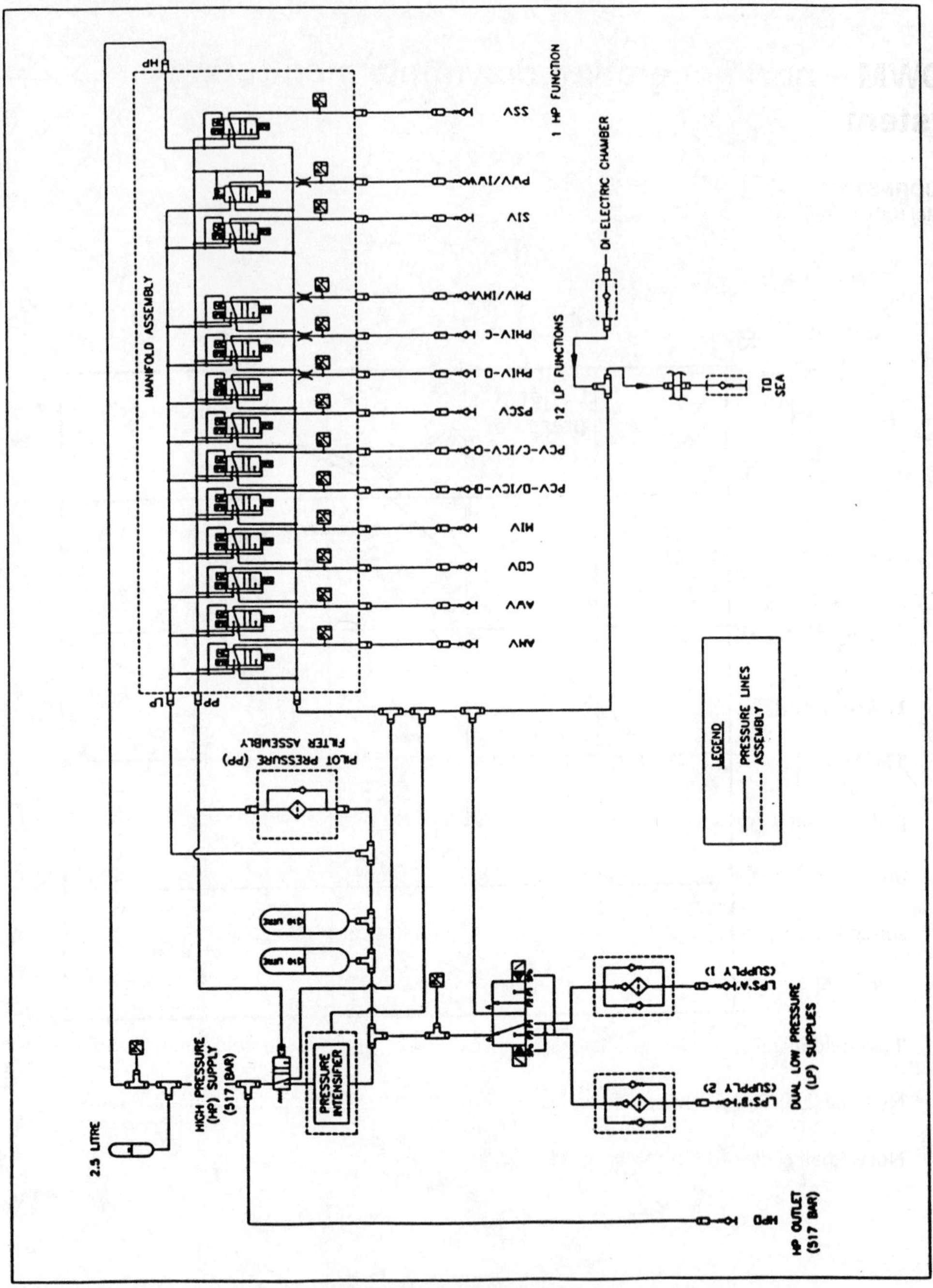


**FIGURE 7 — TYPICAL CONTROL POD HYDRAULIC SYSTEM INCORPORATING AN HPI**

# ꞏOWM – new generation downhole monitoring ꞏystem

**BJORNSTAD**
ꞏcatel Kabel Norge as

**1. ABSTRACT**

The Fibre Optic Well Monitoring (FOWM) system is available for permanent installation in platform and subsea wells for reliable pressure and temperature acquisition. FOWM is based on optical sensors in combination with fibre optic communication links, and is especially well suited high pressure, high temperature (HPHT) applications.

Two field installations have so far been made, namely in the Sleen field in the Netherlands for Nederlandse Aardolie Maatschappij B.V. (NAM), and offshore in the Gyda field in the Norwegian sector for BP Norway.

This paper will review the FOWM technology and present the experience from the field installations.

## 2. INTRODUCTION

On-line downhole pressure and temperature data is used to optimise hydrocarbon production. Permanently installed electrical downhole monitoring (DHM) systems are used more frequently to provide this data in a cost efficient manner. The reliability of these systems have historically not been adequate, and not in line with the target specification currently being considered as a practical aim of 90% probability to perform within the system specification over a period of 5 years [1].

Further challenges to the reliability of DHM systems are:

- Future marginal field developments are planned with few or no workovers, i.e. the DHM system life needs to be equivalent to field life of 10 to 20 years.
- New applications in deeper reservoirs will have higher pressure and higher temperature characteristics.

Future high temperatures applications impose a severe problem on the use of electrical DHM systems. Only special "MIL-spec" components can be used for applications up to 120°C continuously and up to 150°C on short term bases.

The reliability of permanently installed electrical DHM systems is insufficient for current applications and will be further reduced for future applications with high pressure and temperature characteristics and longer periods between interventions. In that light, a new and diverse technology approach to the problem is made to realise the high reliability targets.

## 3. FOWM

FOWM is a new type of permanently installed DHM system [2]. In contrast to the currently available electrical systems, the FOWM system is based on an optical sensor system concept in combination with optical communication [3].

The high reliability requirement for permanent installations, and the anticipated HPHT applications are covered in this concept, which is reflected in the main design parameters:

- Pressure range: 1 - 700 bar
- Temperature range: 20 - 170°C
- Useful life: ≥ 5 years, 90% probability

Main elements that contribute to the high reliability potential of the system include:

- Reduction of the number of downhole components; only passive pressure and temperature sensors are situated downhole.
- Complete elimination of downhole electronics; all required optoelectronics are platform based (alternatively in subsea pod), simplifying access for possible maintenance.
- Exploitation of a fibre optic communication channel, insensitive to electrical noise and the equivalent of electrical short circuit failure mode.
- Slim cable design with laser welded steel tubes of "unlimited length", minimising the risk of cable damage during installation and operation.

The sensor element comprises a micro-machined silicon oscillator which is activated and read optically through the same fibre. The resonant frequency of the oscillator is determined by the pressure and temperature downhole. The operating principle is practically insensitive to fluctuations in signal attenuation, and allows for a large separation in distance, more than 25 km, between the gauge and the acquisition unit.

The reliability requirement for permanent installations has been a major design criterion from the start of the project; the system components have been designed accordingly.

The FOWM gauge contains one pressure sensor and one temperature sensor, and has a slim design measuring only 25 mm in diameter, which is an advantage especially in slim bore wells.

The laser welding technique utilised in the steel tubes containing the fibres in the downhole cable enable manufacture of "unlimited" lengths of cable without factory splices. The current standard cable length is 10 km. This is a great advantage considering that electrical cables typically fail in the factory splices.

The optical fibre penetrator uses slim 3.2 mm (1/8") tubes, and serves as pressure barriers down to the fibre level in the wellhead.

The acquisition unit comprises two modules, the Conversion and Logging Unit (CLU) which controls the system and presents and logs pressure and temperature data, and the Optoelectronics unit (OE) with the core optical and electronic components which has been realised on only 3 standard Europe-cards, and can be mounted in control pod or rack. The system has standard RS232 interface and can be integrated with control systems, or come as a stand-alone unit with data displayed on a rack mountable PC.

Subsea completions require wet mateable connectors, and units for both wellhead and pod applications have been developed and tested [4]. The pod connector unit is also available in ROV/diver operated version.

## 4. SYSTEM CHARACTERISTICS

The gauge has a simple frequency response both with respect to pressure and temperature loads, which makes it easy to implement the conversion algorithms in software. The pressure response is typically 45 Hz/bar and the temperature response is typically 15 Hz/°C.

The basic gauge resolution is less than 0.5 mbar while the resolution of the complete system, which is determined by the resonance peak frequency resolution of the optoelectronics, is less than 5 mbar. FOWM has a time response of 1 bar/s, i.e. can follow such pressure changes and stay within specifications.

The system has a large degree of flexibility, which covers the user's varying needs during the life of the well. The user can choose logging interval of 2.3 seconds or any multiple thereof in order to get proper data coverage during pressure transients.

The accuracy of the system has been studied by testing FOWM in parallel with highly accurate reference sensors, and has been observed to be within ±0.05 bar.

The stability of the system has been investigated during a 45 days test at 500 bar and 170°C. Minor variations of less than ±0.2 bar occurred due to the accuracy of the measurement set-up (Reference pressure sensor at room temperature), but no drift trend has been observed.

## 5. SLEEN 7 INSTALLATION

The first system was successfully installed in an onshore gas well in the Sleen field in the Netherlands for NAM in October 1993.

The Sleen 7 downhole conditions of 120 bar and 82°C are relatively mild, and, in that respect, the well does not reflect the HPHT capability of the FOWM system.

Intermediate measurements were made during installation with an Optical Time Domain Reflectometer (OTDR) to check the integrity of the downhole cable. At the same time status of the gauge was tested with the FOWM acquisition unit. No degradation was observed. The system became operational as soon as the last splice was made.

The system at Sleen is supplied with a modem for remote monitoring. NAM and Shell Research/Amsterdam are monitoring the well from their offices in Schoonebeek and Amsterdam, while Alcatel Kabel Norge can perform diagnostics and service from Norway. System availability is important for remote operation, and automatic re-boot of the system after power failures is implemented.

A well test was conducted in November 1993 following the installation, comprising a well flowing element and a pressure build-up test. Downhole pressure and temperature information was acquired with the FOWM system and with a temporally installed memory gauge. FOWM data was acquired and stored remotely by NAM and KSLA by using the modem facility.

The system dynamic response is an important parameter to be quantified, i.e. the pressure changes that can be followed by the system without loosing the specified measurement accuracy. This parameter is most relevant during the initial transient period of a pressure build-up test. In this test information on the slope and duration of the pressure increase after well closure is acquired, which is indicative for reservoir permeability and for well in-flow performance. An accurate description of these effects is used to tune and optimise production.

The second part of the well test programme comprised a pressure build-up test. The well was closed after being on stream at maximum rate, and the pressure build-up test lasted for several days.

During the pressure build-up test both the FOWM system and a memory gauge were used. The memory gauge was sampling once every 7.5 s as below:

- Accuracy: 0.33 bar
- Resolution: 0.001 bar

The FOWM system was acquiring data once every 2.3 s as specified below:

- Accuracy: 0.2 bar
- Resolution: 0.04 bar

Results as obtained with the two systems show an identical general trend for pressure build-up, although the FOWM system showed somewhat higher pressure derivative probably due to better time response. The fastest pressure change observed was 0.05 bar/s.

Comparison of the two data sets obtained during this week test period demonstrated that the FOWM system was performing according to specification, also after being installed in a producing well.

In order to assess the long term stability of the FOWM system further wireline tests were planned. On June 1st 1994, more than 7 moths after the installation, a Static Pressure Gradient survey (SPG) was conducted at the Sleen 7 well.

During the SPG survey a wireline tool was lowered in the non-producing well to acquire pressure data at various depths. Two Amarada gauges and a thermometer device were contained in the wireline tool. The tool was calibrated with a Dead Weight Test (DWT) reference system before entering the well.

Quoted specifications for the wireline measurement are listed below:

- Accuracy: 0.3 bar
- Resolution: 0.075 bar

Pressure measurements were taken at various depths with the two Amarada gauges, and the maximum temperature during the total wireline operation recorded.

Results presented below are matching within the accuracy and resolution ranges of the two systems. The presented results indicate that the long term stability of the FOWM system is good after being operational for more than 7 months. The results for the temperature measurement also agree well and are within specification.

| System | Pressure (Bar) | Temperature (°C) |
|---|---|---|
| Wireline | 92.94 | 82 |
| FOWM | 93.09 | 82.23 |

*FOWM stability test 7 months after installation, comparison with wireline results from static pressure gradient survey at Sleen 7*

A further interesting quantitative observation of the FOWM temperature reading is the stability in this reading over the 7 months. Downhole temperature conditions can be assumed as rather constant over such period in a closed-in and in a producing well condition. Mean temperature and related standard deviation measured for both closed-in and the producing well status have been calculated and are presented in the table above. The deviation is 0.16% for closed-in conditions, and 0.24% for producing conditions.

The information in the table below provides a second and independent indication of the stability of the FOWM temperature sensor. It should be noted that percent deviation includes both measurement error and also fluctuations in downhole temperature. The larger standard deviation calculated for the producing well status could be indicative for this effect. With the pressure and temperature sensors having an identical design, the same stability indication as obtained from the temperature sensor can also be assumed for the pressure sensor.

| Well status | Mean temperature (°C) | Standard deviation | % |
|---|---|---|---|
| Closed-in | 82.112 | 0.133 | 0.16 |
| Producing | 83.463 | 0.203 | 0.24 |

*Stability in FOWM temperature reading since system installation at Sleen 7*

The third wireline test was made in March 1995 after 1.5 year in operation. Highly accurate wireline gauges were used with an specified accuracy 0.071 bar. The difference in pressure measurement between the wireline gauge and FOWM was within this stated accuracy, and show that the FOWM system has no significant drift over the period of 1.5 years. The latest update on the system is that it continues to operate after 2.5 years, and that no signs of degradation are observed.

**6. GYDA 2/1-A-32 INSTALLATION**

The first offshore installation of FOWM was performed in the Gyda oil field in the Norwegian sector for BP Norway in August 1994. The Gyda 2/1-A-32 well is characterised as a HPHT well with reservoir condition of approximately 400 bar and 160°C [5].

The FOWM system performed within specifications during perforation and initial well testing. The system was able to follow transients except in two instances during perforation when the pressure change exceeded 1 bar/s, but pressures and temperatures were captured shortly thereafter. During well testing the changes were in the order of 0.5 bar/s and the system sampling interval was set to 2.3 s.

After 3 months in operation fibre breakage developed in the splices between the downhole cable and the penetrator in the tubing hanger. The fusion splice proved to be weak, and the recoating material introduced thermal stresses which caused the fibres to break. This has now been rectified by using high strength splicing techniques which gives the splice equivalent strength to the fibre itself, in combination with recoating materials fully compatible with the thermal expansion characteristics of the glass and coating of the fibre. Even though Gyda is out of operation, it has been a valuable experience in order to establish the necessary technology for using fibre optics downhole.

## 7. GUILLEMOT GUA P2

The first subsea system is now being manufactured, and will be installed for Shell UK Exploration and Production in the Guillemot field in October 1996. Guillemot is part of Shell's Central North Sea FPSO development. This will represent the first field installation of the connector units.

## 8. CONCLUSION

- A switch to fibre optic technology for DHM system is needed in order to meet the reliability requirement of 5 years and probability of 90%, and fibre optics is especially urged by HPHT applications.
- The first permanently installed fibre optic downhole system was installed onshore in October 1993, and is successful both in terms of installation procedures and performance during a 2 1/2 year period. Wireline results from the dynamic well test and the SPG survey have been interpolated and compared with FOWM data. Within the accuracy of the

instruments, results are identical. The comparison shows that the FOWM system is performing well and within specification.

- FOWM was installed offshore for the first time in August 1994. The system functionality and installability onboard a platform and in a deviated well was confirmed.
- Complete fibre optic DHM system packages for platform and subsea installation are available.

## 9. ACKNOWLEDGEMENT

Alcatel Kabel Norge acknowledge:

- The support by Norske Shell as main sponsor of this development, as well as BP Norway, Norsk Hydro and the Norwegian Research Council.
- The contribution from Peter Eigenraam of Shell Research/Amsterdam for post-installation data from the Sleen 7 installation.
- The contribution from Morten Eriksrud of Optoplan for technical support and guidance.

## 10. REFERENCES

[1] B. Bjørnstad, Morten Eriksrud, "Fibre Optic Well Monitoring System", OTC7748, OTC 1995, Houston, 1-4-May 1995

[2] B. Bjørnstad, "Fibre Optic Well Monitoring System", SPE 23147, Offshore Europe, Aberdeen, 3-6 September 1991, pp 425-432

[3] T. Kvisterøy, "Optically excited silicon sensor for permanently installed downhole pressure monitoring application", Sensors and Actuators A, 31 (1992), pp 164-167

[4] A. Berg, "Underwater mateable connector with single-mode fibers", Fiber Optic Components and Reliability, SPIE, Boston, 3-6 September 1991

[5] M. Rødseth, BP Norway (1994), Private communication

# he development of a drive close subsea choke alve actuator for deepwater applications

B MARTIN
ent Introl Limited

## SYNOPSIS

For many years choke valves have been operated using a stepping type actuator on open loop control, with an inherent 'fail last' position characteristic. This paper details the design, development, testing and operational experience of a drive close hydraulically operated subsea choke valve actuator, which could be used on either open or closed loop control. It describes the functional requirements of the choke and actuator unit, reviews the traditional stepping type actuator design and indicates how the philosophy of using a rotary hydraulic motor for choke actuation, evolved. The subsequent testing programme undertaken is detailed and following operational experience, the paper describes how the actuator was further developed with the drive close capability, for a Gulf of Mexico deepwater production choke application.

## FUNCTIONAL REQUIREMENTS

The choke and actuator unit is considered to be one of the most critical components within the production system, for retrieval of hydrocarbon reserves, particularly when the equipment is located subsea. It is essential that the choke and actuator assembly provide long term reliable operation to facilitate optimised recoverable reserves through proactive reservoir management. As illustrated on Figure 1, production chokes are normally located downstream of the production wing valve on the xmas tree. Their function is to provide both pressure and flow control, by means of a variable area fluid control device. Figure 2 illustrates a typical subsea choke design, whereby the plug head moves in a linear direction to vary the developed flowing area in the guide component and provide the necessary control characteristics. In undertaking this fluid control, the choke is required to dissipate energy by converting from the fluid pressure phase into the velocity phase i.e. potential energy into kinetic energy. Some of the consequences of this energy dissipation, particularly on high pressure drop applications, can be that the choke is subjected to potential excessive vibration, high levels of noise generation, cavitation, erosion and even premature failure. During the sizing/selection procedures the energy dissipation levels are taken into account in order to minimise their effects. The Kent Introl technology that supports the various subjects reviewed during the sizing/selection process is based on many years of handling fluid control devices on both control and choke valve applications.

The choke actuator is an integral part of the complete assembly, used to position the choke to a predetermined setting, depending upon the fluid control requirements. The actuator is also required to maintain stable control of the choke under all operating conditions. Successful operation of the actuator is therefore an important factor relative to the overall performance of the choke valve assembly. Further requirements of the actuator include the provision of manual/ROV override facility, together with local and remote position feedback signals, to ensure that the choke is providing the required control characteristics.

Actuator sizing is an important aspect of the selection process to ensure that it is capable of fulfilling its functional requirements under all operating conditions. Thrust calculations are carried out with the choke in the closed position, operating against the maximum pressure differential, which is normally assumed to be the worst case, i.e. with full shut-in pressure upstream of the choke and zero downstream. It is also necessary to consider thrust requirements for all the operating conditions, including choke start-up and when there is a possibility of the full working pressure being 'locked in' through closure of upstream and downstream isolation valves. Consideration also needs to be given to other effects within the choke valve including stem/plug seal friction and dynamic effects resulting from fluid flow. The actuator is required to be sized to handle these thrust requirements at the minimum specified hydraulic fluid pressure and also capable of withstanding the forces generated when the unit is operated at maximum hydraulic fluid pressure.

Other factors to be considered during the actuator selection phase include the type of hydraulic fluid to be used, hydraulic fluid pressure available, operating water depth of the actuator and materials of construction. Most actuators are capable of operating with both water based and mineral oil type hydraulic fluids. Individual considerations are required for each of these fluids with respect to performance characteristic, material selection, lubrication compatibility and recirculation. With regards to the operating depth of the actuator, this also needs to be considered in the overall design. Some of the factors that require review include the need to consider pressure compensation of the actuator and any additional forces necessary to return the hydraulic fluid within the control system. This is particularly important when the hydraulic reservoir is located on a topside facility and the hydraulic fluid is returned from subsea components. In addition to the normal considerations of material selection for a mechanical assembly, it is necessary to take into account other factors. These include compatibility with the operating and lubrication fluids, together with consideration of the environment into which the unit will be operating i.e. subsea. In addition to general corrosion effects it is necessary to consider other corrosive mechanisms such as galvanic and crevice corrosion.

## TRADITIONAL CHOKE ACTUATION

As previously indicated, the traditional method of actuating automatic subsea choke valves has been to use an hydraulic stepping actuator. A typical operating philosophy for this type of actuating mechanism is illustrated on Figure 3. Following a signal from the control system, either the open or close actuator hydraulic cylinder would be activated and initiate movement of the respective cylinder drive fork onto the starwheel. Full movement of the drive fork would rotate the gear reduction unit wormshaft and wormwheel simultaneously. Rotary movement of the worm and wormwheel arrangement would be converted to linear movement of the choke stem through the use of a high duty Acme thread and anti-rotation device. The actuator hydraulic cylinder would then be depressurised and return to its start position, assisted by the internal spring arrangement, thereby completing a single pulse movement. The actuator operating sequence could then be repeated using the identical hydraulic cylinder to move in the same direction or alternatively using the second hydraulic cylinder to move in the opposite direction. Between hydraulic pulses, the actuator would be locked in position by the irreversibility of the Acme thread, and the actuator inherently fails in its last position on loss of hydraulic power or electrical signal failure. These types of actuators could be fitted with an ROV override facility, local and remote position feedback devices.

The actuator could also be required to be utilised as part of a subsea retrieval insert philosophy, whereby the actuator and trim internals would be changed out subsea, using either diver assisted, ROT diverless or ROV diverless intervention systems. Figure 4 illustrates a typical subsea choke valve and stepping hydraulic actuator assembly supplied for a deepwater application, whereby the actuator and trim cartridge assembly could be replaced subsea, using ROV diverless intervention techniques. This particular design of actuator has a typical opening time of 10 minutes for operating the choke from the fully closed to fully open position or vice versa. Other actuators, using the stepping principle can require several hours to operate the choke over their full travel.

With stepping type actuators, choke control is normally achieved using an open loop philosophy. The control system sends a number of predetermined hydraulic pulses to the actuator in order to position the choke at the required opening. Choke position is determined either by use of a remote position feedback device or alternatively counting and recording the number of pulses the actuator has been operated.

## ROTARY ACTUATOR DEVELOPMENT

During 1987/8 Kent Introl participated in a joint development project with BP Petroleum Developments to further develop a subsea choke and actuator unit for use in conjunction with their Diverless Subsea Production Systems (DISPS) philosophy. As part of the development, the choke valve actuator was identified as one of the critical factors in the overall production system. The DISPS project was divided into three phases, these being:-

(i) Conceptual design study
(ii) Detailed design study
(iii) Testing

As part of the detailed design phase, an analysis of current designs and existing marketed products was undertaken. The objectives derived from this market analysis indicated that the actuator should be simple, robust, extremely reliable and capable of being operated using existing choke control philosophy. Kent Introl had been developing its own subsea actuator prior to the DISPS project, based on the use of a bi-directional rotary hydraulic drive unit and a gear reduction unit with Acme drive output. This actuator, illustrated on Figure 5 was to be used on a pilot subsea separator system, being developed by B.O.E.T. The conclusion of the marketed product analysis was that the Kent Introl Series F-RTL actuator should be used as the basis for further development during the DISPS project. This was primarily due to the design satisfying more of the desired features than any of the other marketed products reviewed.

Kent Introl had supplied a number of diver operated subsea choke valves including the unit illustrated on Figure 6. This particular choke and actuator was manufactured for Occidental Petroleum (Caledonia) Limited (now Elf Enterprise Caledonia Limited) for use as a Water Injection choke on the Claymore subsea template. The actuator was a diver operated unit, comprising of a worm and wormwheel gear reduction unit and Acme thread output. Actuation was achieved by rotating the input wormshaft, which was converted into linear movement of the choke stem through the Acme thread and anti rotation device. In order to provide a much simpler automatic design, the principle of using an hydraulic motor drive unit in conjunction with the worm and wormwheel gear unit was evolved. On a similar basis to the Claymore water injection choke, the local position monitoring and required remote position feedback devices would be driven direct from the choke/actuator stem. This design philosophy, illustrated on Figure 7, resulted in an economic, compact, robust design with significant size and weight reduction, in comparison to the original Series F-RTL unit.

The original Series F-RTL unit had been developed using a Moog-Donzelli axial piston hydraulic motor and during the DISPS project a review of a number of hydraulic motor drive unit types was undertaken. Three types of motors were considered for the application. The criteria reviewed for each type included principle of operation, materials of construction, number of components, low speed/low flow characteristics, case drain requirements, maintenance, fluid consumption and suitability for subsea choke application. The three motors reviewed were:-

(i) Axial piston motor - Moog Donzelli Series A-084
(ii) Radial piston motor - Mannesmann Rexroth type MCR05 Series
(iii) Geared motor - Adan Type AWM

The unit determined to be the most suitable and which satisfied the majority of the requirements was the geared motor. A more detailed test programme was initiated using this drive unit in conjunction with the worm and wormwheel arrangement previously described.

## ACTUATOR TESTING

Following completion of the detailed design phase of the DISPS project, a prototype actuator unit was manufactured in order to further evaluate its performance characteristics and undertake extensive endurance testing. In order to undertake the testing, a special actuator test rig was built as illustrated on Figure 8. This test rig provided the capability of testing the prototype actuator at loads up to 20000 kgf thrust over a travel of up to 60mm, at various hydraulic fluid pressures and in both directions. This rig was also used to carry out the endurance cycle testing. The test programme was divided into testing of the Adan hydraulic motor independently and then the complete actuator assembly. Tests carried out on the Adan geared motor included torque capability, stall characteristics, linearity evaluation, seawater immersion testing and corrosion resistance determination. With the motor assembled onto the actuator, the unit was mounted onto the test rig and performance/endurance testing carried out. This included tests at various actuator loads between 2000 and 20000 kgf over 60mm travel, in both directions and also at hydraulic pressures between 100 Bar and 200 Bar. The testing undertaken included functional, performance and endurance aspects.

## OPERATIONAL EXPERIENCE AND DRIVE CLOSE PHILOSOPHY ESTABLISHMENT

Following the DISPS project, Kent Introl supplied a number of subsea chokes incorporating the Series F-RTL actuator. Figure 9 illustrates an example supplied to Arco British for use on the Welland subsea development system. Three of these chokes were supplied in 1989 and have been operating satisfactorily, since their installation. Similar units have been supplied on other subsea projects and all those supplied in the North Sea to date, have incorporated conventional fail last position characteristics.

During the DISPS project it was recognised that the actuator developed, could be used either to fail in its last position on loss of hydraulic supply or alternatively be driven to a fail position. BP Exploration had considered the development of a separate fail close valve for use in the Gulf of Mexico. This was required as part of the procedure for testing the xmas tree master valve, a test which has to be undertaken at 6 monthly intervals in order to comply with operational regulations. The procedure would be to close down the xmas tree, depressurise above the master valve and test for leakage between the master and production wing valves. It was recognised that during closure of the production system, components not normally used to regulate fluids could be subject to potential damage due to high levels of energy dissipation. A philosophy of using the production choke with a high duty trim configuration, to provide flow closure, was therefore proposed, including the requirement for the actuator to have a drive close capability. It was recognised that the choke would require to have a subsea retrievable insert configuration and therefore, if damage was to occur, the choke insert could be replaced. In addition to its normal function the choke therefore had a second requirement, which was to act as a 'flow cut' valve. This eliminated the requirement for an additional valve to undertake this function.

## BP POMPANO PHASE II PROJECT

Following discussions between BP Exploration and Kent Introl, the objectives for the Pompano Phase II subsea development production chokes were defined, as follows.

(I) Development of an insert type ROV retrievable choke using proven subsea components.

(II) Provide an actuator with a continuous drive close hydraulic motor.

(III) Actuator and choke to be designed to interface with the Pompano Phase II Xmas tree and control system.

(IV) Schedule to be consistent with the Pompano Phase II project requirements.

Using the actuator design developed during the DISPS project and subsequently used on a number of North Sea subsea projects, it was considered that the existing configuration would not be ideal for ROV intervention. It was therefore proposed that the existing design would need to be reconfigured to adapt the unit for ROV intervention and also provide the drive close capability. Kent Introl had supplied a significant number of chokes having subsea retrievable actuator and trim cartridge inserts. These included diver assisted types for shallow water and ROT/ROV diverless units, for deeper water subsea applications. The choke body to insert connection system had used various mechanisms, including a screwed locking ring for the diver retrievable units and collet arrangement for the ROT diverless design. On the Pompano Phase II project, ABB Vetco Gray, an associate company of Kent Introl, was involved in the supply of the subsea diverter valves. These particular valves utilised subsea retrievable inserts, based on a ROV diverless system. The valve body to insert connection system of the diverter valve used a screwed locking ring, similar to the type used by Kent Introl and the ABB Vetco Gray scope of supply included provision of the running/retrieval tool. It was, therefore, proposed that the choke would be configured to use the same running/retrieval tool as the diverter valve, thereby standardising on the subsea retrieval philosophy for these two components. As illustrated on Figure 10, this resulted in the development of an insert type ROV retrievable choke based on components with proven subsea experience.

On the basis of the requirement to provide a continuous drive close hydraulic actuator, the philosophy developed during the DISPS project had been proposed. The actuator could be operated using either a series of hydraulic pulses or alternatively providing a continuous hydraulic signal to the unit, which would drive the actuator in a specific direction. In the event of loss of hydraulic pressure it was proposed that an accumulator would provide the necessary fluid source for full operation of the choke under all conditions. In order to adapt the existing series F-RTL design to be more suitable for ROV intervention, it was agreed that the unit would need to be reconfigured. Using the same operating philosophy as the original DISPS actuator, the design illustrated in Figure 11 was formulated. The most significant modification was replacement of the worm and wormwheel gear reduction unit with a bevel gear arrangement. Following initial testing of the alternative gear reduction unit in conjunction with the Adan motor, a full prototype actuator was manufactured.

An extensive prototype test programme was initiated based on that undertaken during the original DISPS project. This testing included the following:

(I) Power output relative to hydraulic fluid pressure.
(II) Stall characteristics relative to hydraulic fluid pressure.
(III) Operational characteristics relative to pulse length and hydraulic fluid pressure.
(IV) ROV override torque relative to load.
(V) Local and remote position mechanism accuracy and repeatability.
(VI) Actuator drive close capability.
(VII) Actuator endurance testing.
(VIII) Actuator hyperbaric testing.
(IX) Actuator insert stab test.

Following completion of the test programme, BP Exploration placed an order on Kent Introl Ltd, for the supply of 10 production choke and actuator units.

As indicated previously, chokes have traditionally been operated using a stepping type principle whereby the choke position is determined by counting the number of pulses. Following the extensive testing carried out on the prototype and discussions with GEC Marconi, the control system manufacturers, it was proposed that the units would be pulsed at 1/4 second intervals with a 4 second delay, to maximise control accuracy. The position feedback would be taken from the remote electrical position device installed on each of the actuators. This effectively was providing a closed loop control for the choke in comparison to the traditional open loop philosophy. In order to provide the drive close capability GEC Marconi provided a bank of accumulators with a continuous trickle charge, within the control pod and the control system was configured to provide full operation of the choke under all operating conditions with loss of the main hydraulic fluid source.

Significant discussions were held between all organisations involved in the project in order to co-ordinate the various activities. The choke and actuator unit required co-ordinating activities of the main BP project team with the xmas tree manufacturer Cooper Oil Tool (USA), controls system manufacturer GEC Marconi (USA and UK), running/retrieval tool manufacturer ABB Vetco Gray (USA) and choke/actuator manufacturer Kent Introl (USA and UK). In order to meet the overall Pompano Phase II project requirements, each organisation had to operate within a strict production schedule. With regards to the choke and actuator units, Figure 12 details the major aspects covered within the overall project from initial conception through to delivery of the 10 production units.

## SUMMARY

This paper described how the requirements for a drive close subsea choke valve actuator was developed for a particular Gulf of Mexico deepwater application. It details how through the use of components and assemblies with previous subsea experience, the final solution was evolved. It also demonstrates how through joint ventures between major oil companies and manufacturing organisations, successful projects can be completed, even within the most stringent timeframes.

A major aspect which led to the success of the project was that the traditional customer and client philosophy was not utilised. This was replaced by a project team philosophy, comprising of all interested organisations, including the operator, project engineering support team, xmas tree supplier, controls system supplier, running/retrieval tool manufacturer and choke/actuator supplier. Within this project team philosophy, discipline teams were organised to discuss and agree individual project requirements. These individual requirements were then integrated into the overall project plan. This project philosophy was applied successfully to the choke/actuator assembly and also to other equipment supplied on the Pompano Phase II project. It is, therefore, a philosophy which should be considered on similar future projects of this type.

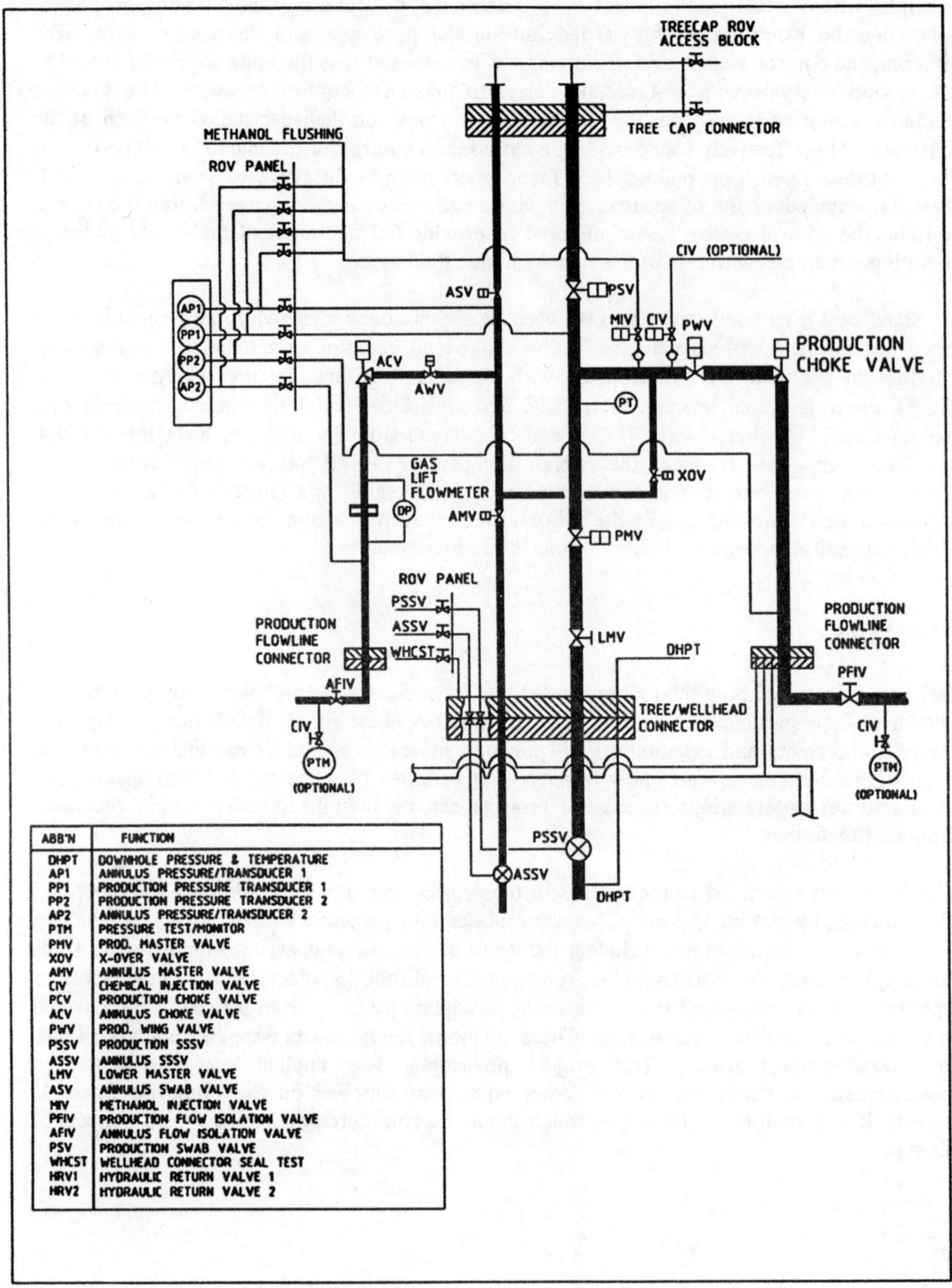


| ABB'N | FUNCTION |
|---|---|
| DHPT | DOWNHOLE PRESSURE & TEMPERATURE |
| AP1 | ANNULUS PRESSURE/TRANSDUCER 1 |
| PP1 | PRODUCTION PRESSURE TRANSDUCER 1 |
| PP2 | PRODUCTION PRESSURE TRANSDUCER 2 |
| AP2 | ANNULUS PRESSURE/TRANSDUCER 2 |
| PTM | PRESSURE TEST/MONITOR |
| PMV | PROD. MASTER VALVE |
| XOV | X-OVER VALVE |
| AMV | ANNULUS MASTER VALVE |
| CIV | CHEMICAL INJECTION VALVE |
| PCV | PRODUCTION CHOKE VALVE |
| ACV | ANNULUS CHOKE VALVE |
| PWV | PROD. WING VALVE |
| PSSV | PRODUCTION SSSV |
| ASSV | ANNULUS SSSV |
| LMV | LOWER MASTER VALVE |
| ASV | ANNULUS SWAB VALVE |
| MIV | METHANOL INJECTION VALVE |
| PFIV | PRODUCTION FLOW ISOLATION VALVE |
| AFIV | ANNULUS FLOW ISOLATION VALVE |
| PSV | PRODUCTION SWAB VALVE |
| WHCST | WELLHEAD CONNECTOR SEAL TEST |
| HRV1 | HYDRAULIC RETURN VALVE 1 |
| HRV2 | HYDRAULIC RETURN VALVE 2 |

Fig 1 Location of Production Choke on Xmas Tree

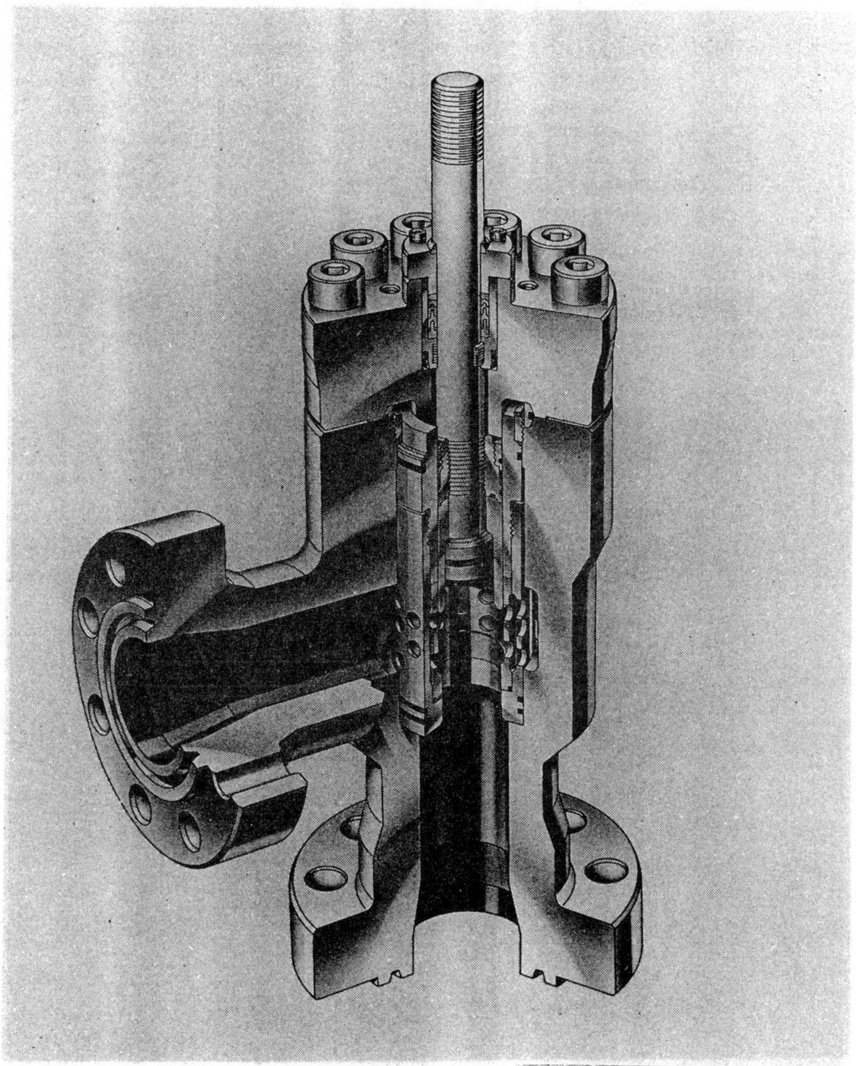

Fig 2 Typical Subsea Choke Valve design

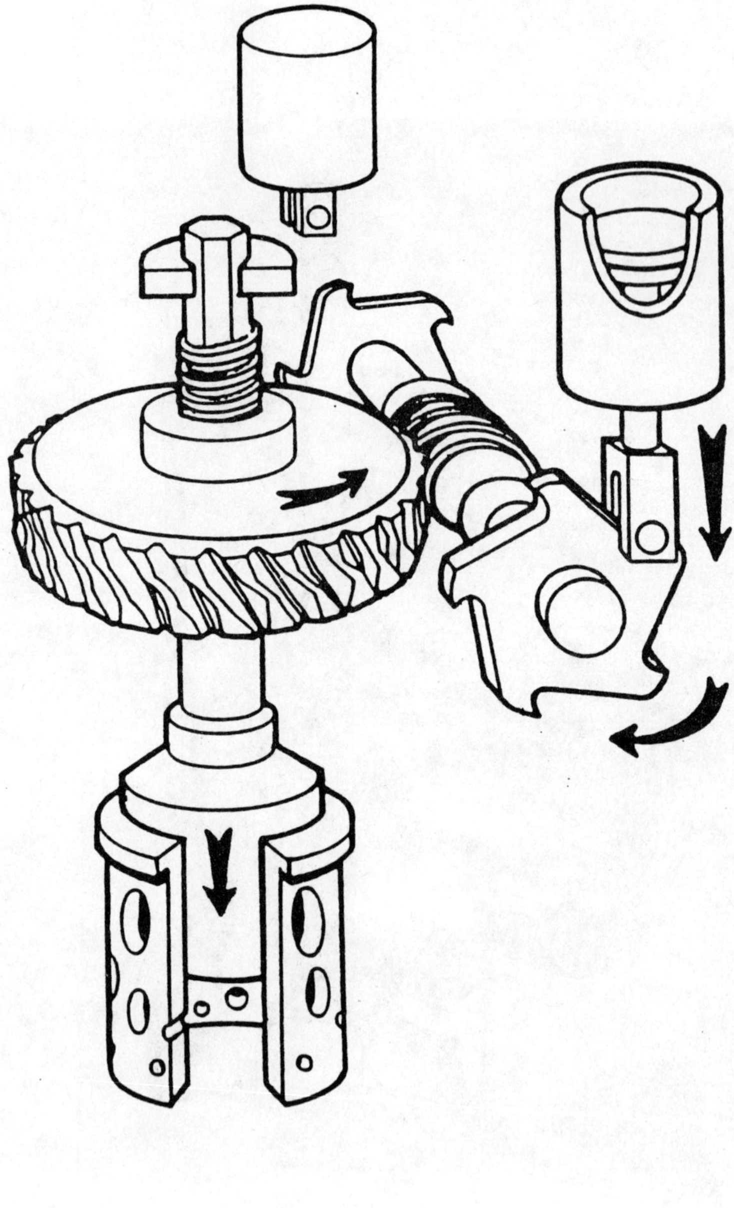

Fig 3    Hydraulic Stepping Actuator operating philosophy

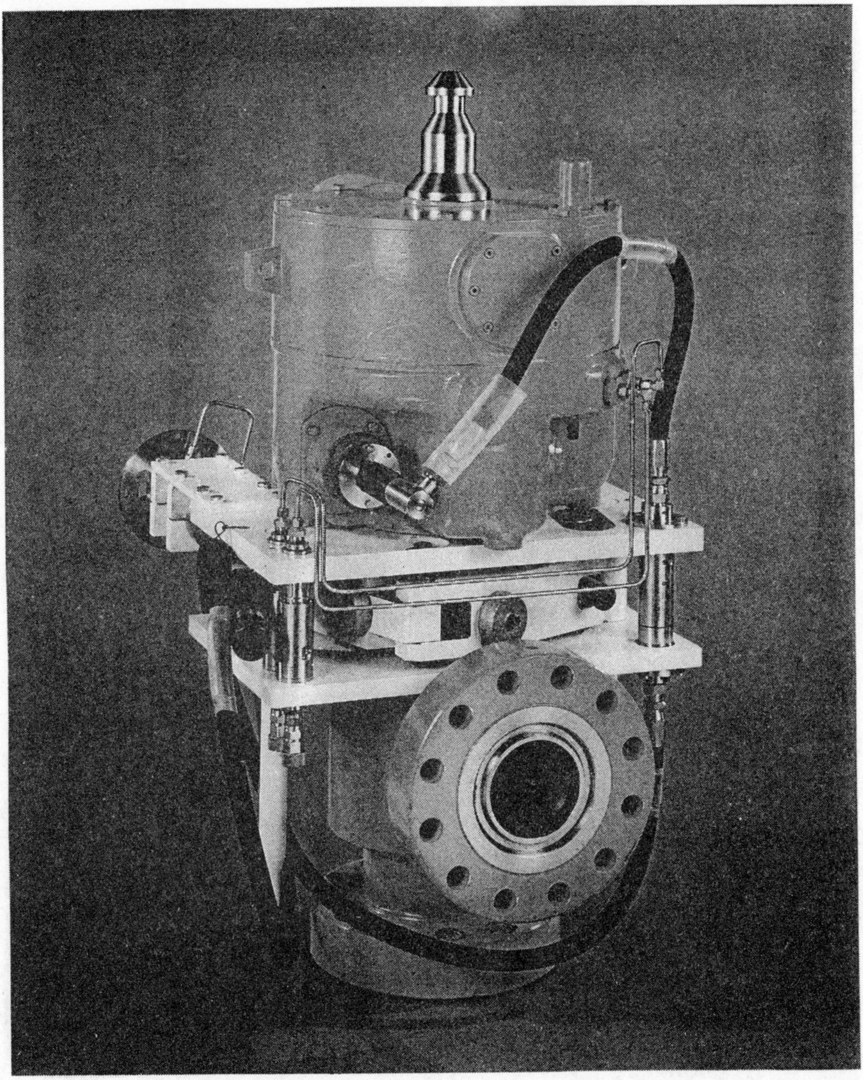

Fig 4 Typical Subsea Choke Valve and Actuator assembly suitable for ROV diverless intervention systems

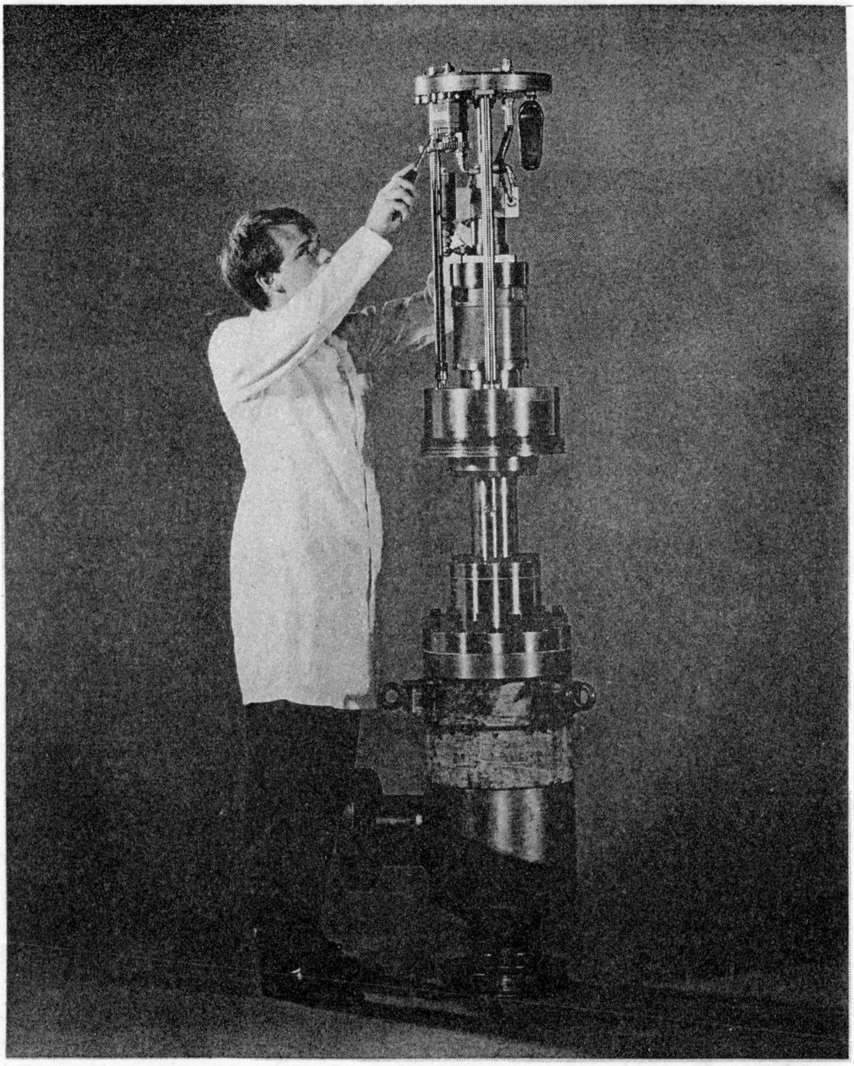

Fig 5  B.O.E.T. Subsea Hydraulic Actuator

Fig 6 Occidental Petroleum/Claymore Subsea Template Water Injection Choke

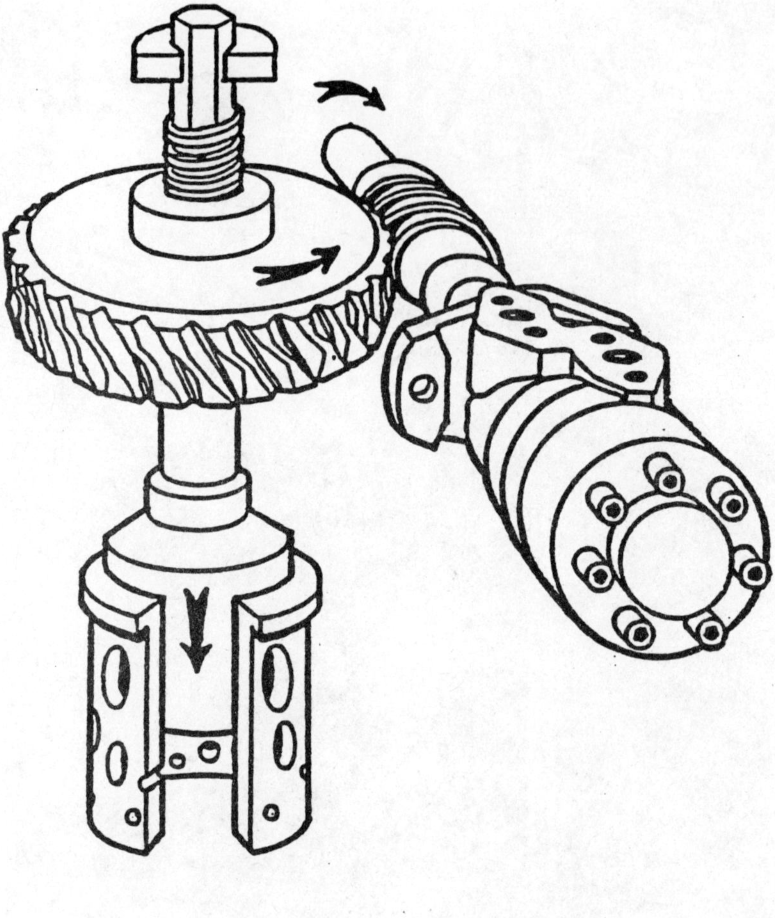

Fig 7  D.I.S.P.S. Rotary Hydraulic Actuator operating philosophy

Fig 8 D I S P S Actuator test facility

Fig 9 Arco British Welland field Subsea Choke Valve and Series F-RTL Hydraulic Actuator

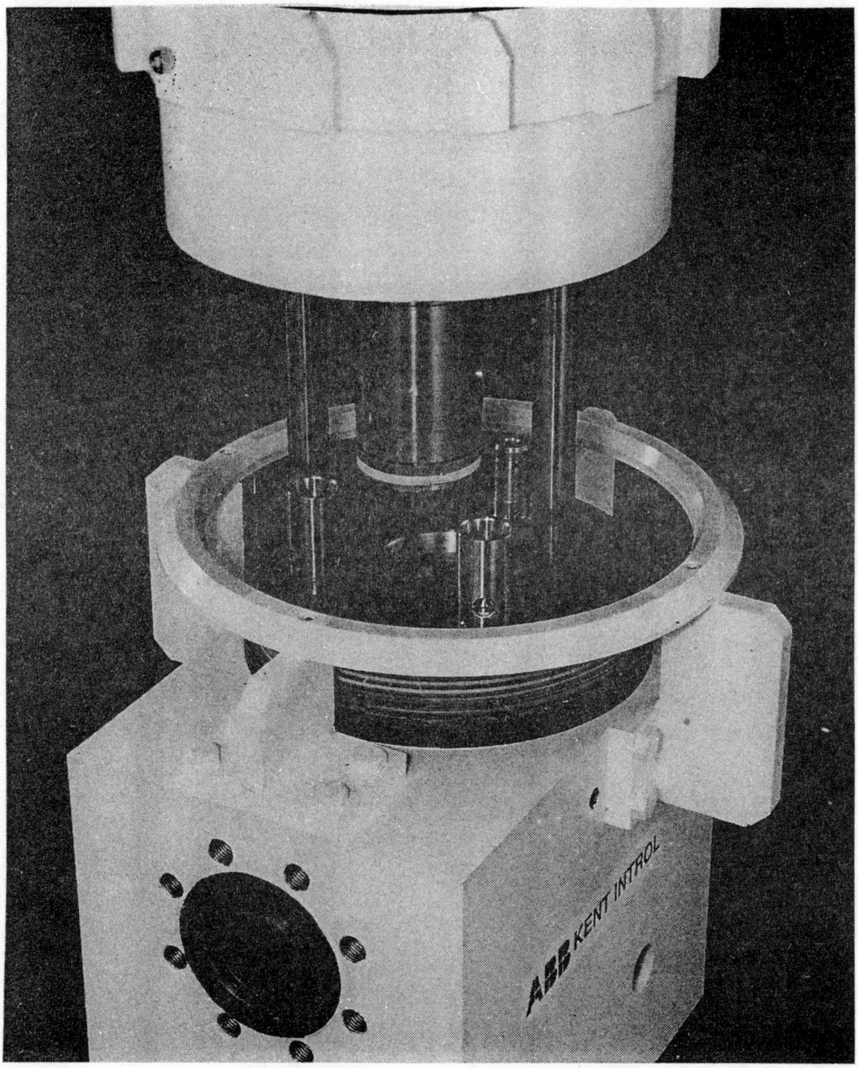


Fig 10 BP Pompano Phase II Insert type Choke design.

Fig 11 BP Pompano Phase II Choke and Actuator assemblies

**BP POMPANO PHASE II CHOKE AND ACTUATOR SCHEDULE**

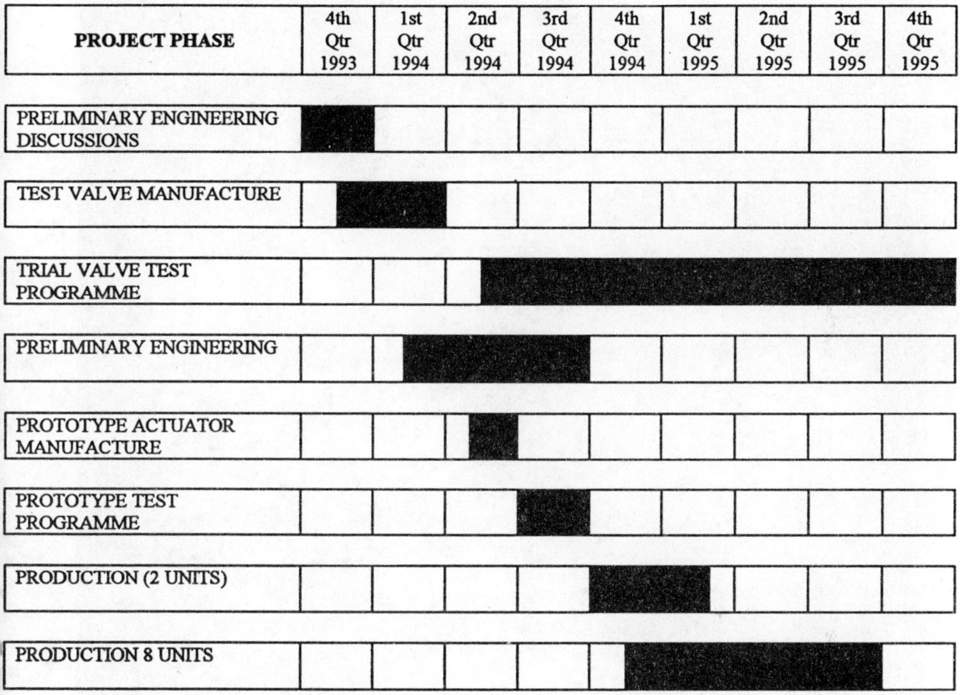


Fig 12 BP Pompano Phase II Choke/Actuator project schedule

# Acoustic Technology

# A non-intrusive sand detection system for subsea nstallation

G FOSSUM and O HELGESEN
mrad Norge as


## SYNOPSIS

A new instrument for detection of sand in multiphase flows subsea is described. Sand particles transported by the flow, induce acoustic 'noise' in the pipeline wall. This 'noise' is detected by a sensor clamped to the outside of the pipeline.

Simrad developed their first acoustic non-intrusive sand detector system for subsea use in 1989. This system was designed for detection of sand in gas pipelines subsea.

There is an increased awareness of the requirement to monitor sand production because many existing fields and new fields being developed are prone to sand production.
The increasing number of subsea wells also makes subsea installed sand detectors more attractive, since the only way to directly detect which well is producing sand is to install a detector on each production pipeline.

Simrad's new generation of subsea sand detector systems is suitable for both gas and multiphase flows. No modification of the production pipeline is required, and the sensor can be fitted to any pipe diameter. The sensor is normally installed downstream after a bend in the pipeline, and the design is such that it can be adapted to become ROV replaceable.

The sand induced acoustic 'noise' is influenced by the flow conditions and the sand particle size and rate. Using a broadband acoustic sensor and processing the signal over the full frequency range of the sensor, makes it possible to optimise the performance of the system for a particular subsea installation.

The main use for a subsea non-intrusive sand detection system, is to monitor the flow for sudden changes in the sand production caused by e.g. a screen breakdown. However, it is also possible to calibrate the system based on experience data after installation, or earlier based on test-rig trials.

The system provides information about variations in the 'noise' level caused by changes in the flow rate itself. This could provide a useful tool for flow trending and also possibly provide information about the build-up of hydrate plugs in the pipeline.

## 1 INTRODUCTION

The exploitation of oil and gas resources is usually from fields with various types of sandstone formations. The formation may be more or less well consolidated, and sand released and transported with the production flow can cause serious erosion damage in piping, bends and valves. Excessive sand production can also lead to sand bridging and to build-up of sand in the production and transport system.

There is an increased awareness of the requirements for equipment to monitor unwanted sand production, as several operators have experienced problems at fields where sand production was not originally considered to be a problem.

Offshore exploration of oil and gas has gradually moved from shallow fields, to fields at increasing water depths. It is therefore expected that complete underwater production units, often remotely controlled from a nearby platform will be more common in the future. This places increasing demands on the lifetime and safety of components due to the high cost of subsea intervention. In order to ensure the safe operation of components exposed to sand erosion, it is therefore important to have a reliable method of monitoring the sand production.

Continuous real time monitoring of the production flow for sand particles, makes it possible to detect changes in the sand production. Such changes may be due to breakdown or damage of the gravel pack/screen where this is used, or may be caused by sudden changes in the reservoir which leads to an increase in the sand production, (e.g. water breakthrough or collapse of the formation).

By continuous monitoring of the production flow, early detection of sand production is possible. This enables an operator to take corrective action to limit the damage caused by excessive sand production, and by that reducing cost. It is expected that in some cases, instead of shutting down a well, it will be possible to maintain the production but at a lower 'sand free' rate.

Monitoring of the sand production is also expected to be a valuable tool during the initial and early production phases from a field.

Systems for detecting sand in the production flow can either be intrusive, or non-intrusive. For an intrusive system, the pipeline must be penetrated in order to install the sensor. Intrusive detectors are usually based on measuring the change in electrical resistance when a material is eroded away, or on direct or indirect measurement of sand particle impacts by a piezoelectric sensor protruding into the flow.

Simrad's non-intrusive sand detection system has the important advantage that no penetration of the pipeline wall is necessary.

Sand transported with the production flow is detected by an acoustic sensor clamped to the outside wall of the pipeline. The sensor detects the sand by picking up the acoustic 'noise' induced in the pipe wall when sand particles hit the inside wall of the pipeline.

## 2 SIMRAD'S EXPERIENCE

In the spring of 1995, Simrad was awarded a contract to supply a total of 7 non-intrusive acoustic sand detector systems to be installed at the Statoil Norne Field. The contract is with Kongsberg Offshore a.s that has an EPC contract for the subsea system at Norne.

Simrad has been involved with sand detection since 1986/87 when the company was chosen as the industrial partner in the development of a prototype acoustic sand detector system. The initial work was successful, and Simrad was awarded a contract to supply a total of seven combined acoustic sand detector and gas leakage detector systems to the TOGI project. These systems were delivered in 1989. Five of the units were installed at the TOGI deepwater template in the North Sea in 1990, and the systems are still in operation.

The acoustic sand detector system for the TOGI project was developed in co-operation with Norsk Hydro and is believed to be the first of its kind for subsea use. This system was developed primarily for detection of sand particles in gas carrying pipelines, and can detect (70μ - 300μ) sand particle concentrations as low as 0.1ppm (vol) for flow velocities in the range 10 - 20m/s, (1).

Simrad's sand detector system is non-intrusive, i.e. no penetration of the pipeline is required. A broadband acoustic sensor is clamped to the outside of the production pipe close to a bend in the pipeline. The sensor is sufficiently sensitive to pick up the acoustic 'noise' induced in the pipeline wall when sand particles hit the inside of the pipeline.

The sand induced acoustic 'noise' covers a wide frequency range (approximately 200 - 800kHz). A sensor with good sensitivity across this frequency range is therefore required in order to have full flexibility in the subsequent processing of the acoustic signal.

A mathematical model that describes the sand induced acoustic 'noise' as a function of flow parameters such as the gas velocity, condensate flow rate and sand particle concentration was developed for TOGI. After installation and calibration, the system provides accurate quantitative measurements of the current sand production rate (provided the gas velocity is given as an input to the system).

Since the development of the sand detector system for the TOGI project, Simrad has participated in several tests. These tests show that the non-intrusive acoustic detection principle also works well for multiphase flows at low GOR's, and at lower flow velocities than at TOGI.

In 1991/92, Simrad developed a new acoustic sand sensor that is smaller than the original TOGI sensor. This sensor is easier to install than the TOGI sensor, and is held in place by means of a spring loaded bracket. No special preparation of the pipeline is required, apart from removal of the paint layer in a small area where the sensor head is in contact with the pipeline. A subsea version of the sensor that can be installed down to a depth of 500m, and an EX version for topside installation has been developed.

Simrad has participated in two tests in England using the new sensor. A test prototype version of the sand detector system was used during a test at AEA Technology's test facility at Harwell in September 1994, and a test at BP's Wytch Farm in Dorset in June 1995. The test at Wytch Farm was carried out on a 2" test section connected to two live onshore oil wells, and with the sensor installed after a 90° bend in the pipeline.
The tests have been useful in qualifying the new sensor for detection of sand in both gas and multiphase flows.

## 3 PRINCIPLE OF OPERATION

When sand particles transported with a gas or multiphase flow hit the inside wall of a pipeline, acoustic 'noise' is induced in the pipeline wall. This 'noise' is detected by a sensor clamped to the outside wall at a position downstream of a bend in the pipeline.
Simrad's sand detector system is therefore non-intrusive, since there is no need for a penetration in the pipeline wall.

The acoustic 'noise' picked up by the sensor, consists of signals generated by the flow itself and the combination of signals from a multitude of single sand particle impacts with the inner pipe wall. The signal is a broadband 'noise' signal within the frequency range of the sensor (approximately 200 - 800kHz).

As the sand concentration increases, so does the sand induced acoustic 'noise'. An example of this is shown in figure 1 below. The figure shows the relative increase in signal level in dB from the signal level with no sand present in a dry gas flow.

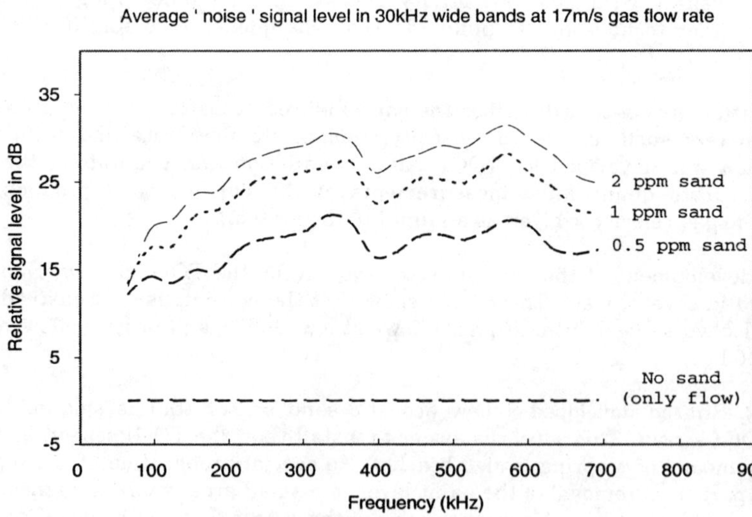


Fig 1 Relative change in signal level for different sand concentrations

The example shown in figure 1, is from a test with the TOGI acoustic sand sensor installed close to a 30° bend in a 6" pipeline, and is for a constant dry gas (nitrogen) velocity of 17m/s. The sand particle mix was 70 - 300micron diameter and the sand concentration is given in ppm (of gas volume).

The acoustic 'noise' signal is converted to two signal levels that represents the current sand production and the background flow induced 'noise'.

Parameters such as flow rate, GOR, sand particle size, flow regime and pipe geometry influence the acoustic 'noise' induced in the pipeline. The kinetic energy of the sand particles is proportional to the mass and the square of the velocity. The velocity at which the sand particles hit the inner wall is the most important parameter and is influenced by the pipe geometry, flow regime and flow rate.

In order to provide quantitative measurements of the sand contents for different flow conditions, it is necessary to consider the parameters that influence the acoustic 'noise'. Such algorithms have not been implemented for the present sand detector system, but can be incorporated provided sufficient data is obtained through realistic testing with known rates of sand for a series of flow conditions.

No special calibration is however required to give an indication of an increase or decrease in the current sand production rate for different flow conditions. The alarm criteria for sand alarms will in this case be based on the rate of change of the sand induced 'noise' signal as well as the absolute level of this signal. Setting an absolute threshold level for the sand induced 'noise' will have to be based on experience after a period of operation, as the absolute signal level is influenced by factors such as the flow rate and GOR.

## 4 SYSTEM DESCRIPTION

The subsea sand detector system consists of the following main modules:

- A sensor fitted in a spring loaded bracket.
- Oil filled jumper cables with penetrator interfaces to sensor, electronics container and ROV connector receptacle.
- A 9 Pin ROV connector receptacle
- A subsea electronics unit fitted in a steel container.

A general arrangement drawing of the sand detector system is shown in Figure 2 below.

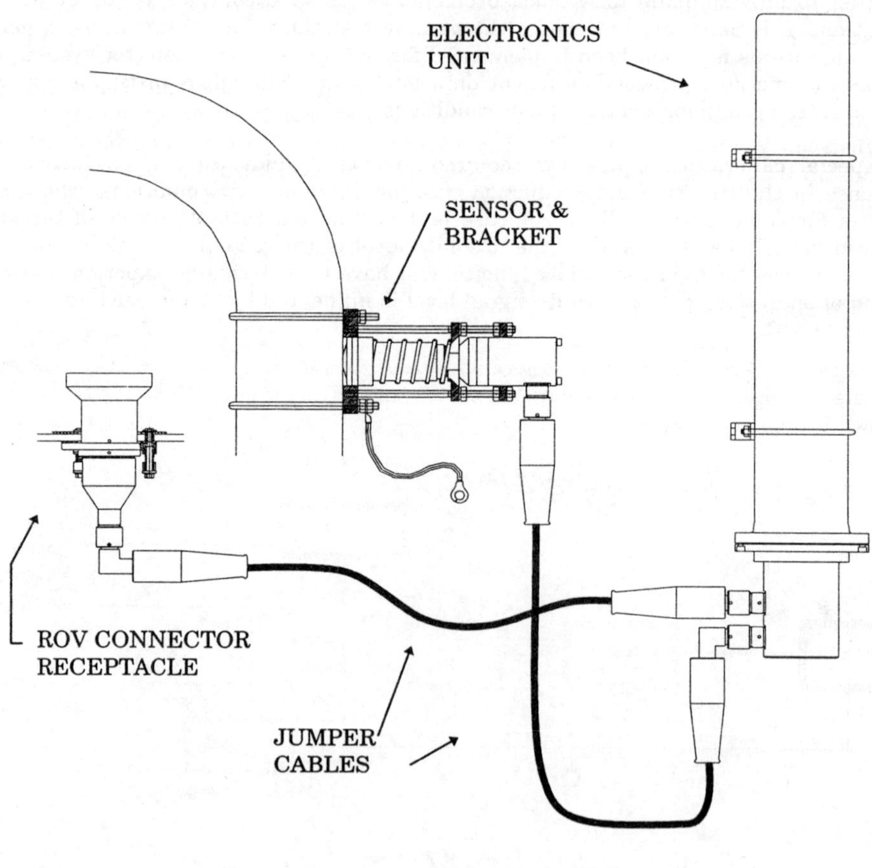


Fig 2 Sand detector system arrangement

## 4.1 Mechanical design

The sensor, electronics container and ROV connector receptacle are connected by means of jumper cables. The oil-filled jumper assembly provides a reliable solution as it reduces the number of underwater connectors in the system to one (nine pin ROV connector receptacle).

The electronics container has an outer diameter of 125mm and a height including the penetrator assembly of 940mm. The container is the same size as the ones used for Simrad's range of long-baseline transponders.

The main parts of the container and sensor are fabricated in a superduplex steel quality. The sensor bracket and some container components are fabricated in SMO 254.

The system is designed for an operational ambient pressure of 50 bar and a test pressure of 75bar.

At Norne, the sand detector system is installed on the choke module. The sensor with bracket is clamped to the outside of the pipeline after a 90° bend in the pipeline and downstream of the choke valve. The electronics container is attached to the structure by means of two U-bolts. The system is interfaced to the KOS control module by means of a jumper cable with a 9pin ROV connector plugged into the plug receptacle (ref. figure 2).

## 4.2 Electrical design

The subsea electronics consists of four single europe format pcb's fitted in a rack inside the electronics housing. A schematic block diagram of the system including the sensor is shown in Figure 3 below.

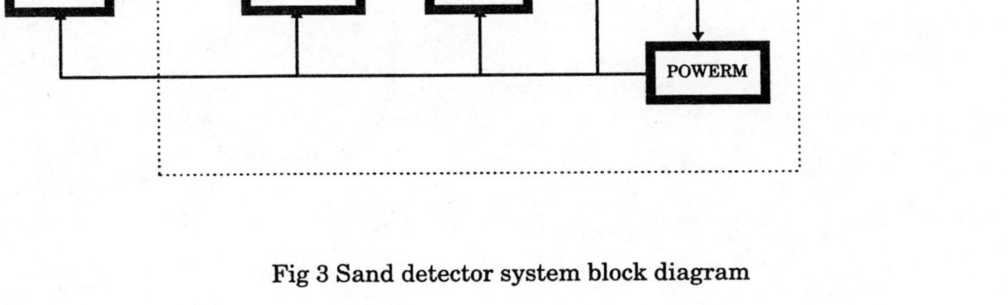


Fig 3 Sand detector system block diagram

SENSOR

The acoustic sensor has an even frequency response and good sensitivity in the range from 200 to 800kHz.

Figure 4 shows the average frequency response of the sand sensor.

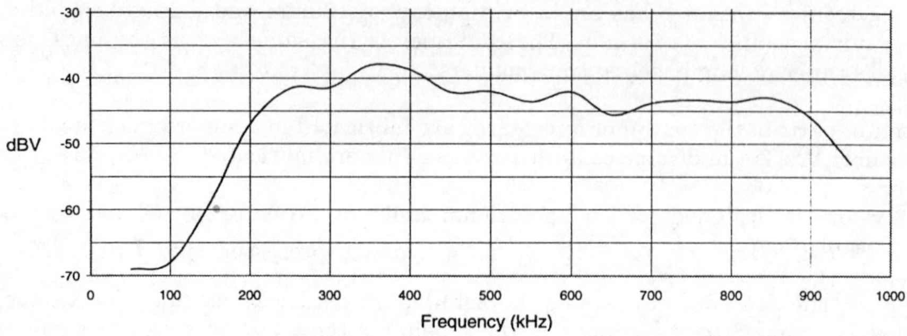


Fig 4 Average frequency response of sand sensor

The sensor has a built-in preamplifier and cable driver which is powered from the electronics unit. Connection to the electronics unit is by means of an oil filled jumper cable. The unit is designed for an ambient operating pressure of 50bar and a test pressure of 75bar.

The sensor picks up the broadband sand induced acoustic 'noise' and the 'noise' from the flow itself and converts the signals to an electrical signal.

INPUT/ADC

The signal from the sensor is connected to the INP/ADC board where it is amplified in a fixed gain/filter stage, followed by an adjustable gain amplifier stage. The signal is converted to digital form and temporary stored in a FIFO memory before being clocked out to the DSPM board.

DSPM

The DSPM board contains a fast digital signal processor (DSP). This processor reads the digitised signal values from the INP/ADC board and performs an FFT that converts the acquired data into the frequency domain. The DSP is also responsible for setting the input gain according to the signal strength of the received signal.

The acquisition process is initiated upon command from the CPUEX board. On receipt of this command, the DSP ensures that the gain is set correctly before digitally sampling the signal from the INP/ADC board. The DSP converts the signal into the frequency domain and outputs the values of the envelope to the CPUEX board for further processing.

CPUEX

The CPUEX board contains the main controlling processor (the CPU). The CPU instructs the DSP to perform data acquisitions at regular intervals. Average 'noise' signal levels representative of the background flow 'noise' and the sand induced acoustic 'noise' are computed. These signals are calculated as the average signal level in two selected frequency bands within the operating range of the sensor. The centre frequency and bandwidth of the bands can be set. This makes it possible to optimise the 'noise' signal level response within two selected frequency bands for a particular installation.

The CPU also calculates the rate-of-change values for the signals from each of the two bands. The value of the signal representative of the sand 'noise' rate-of-change is continuously compared to an alarm limit set by the operator. A sand 'noise' rate of change alarm is set once the signal exceeds the alarm threshold. The CPU also performs internal housekeeping functions and handles all communication with the host computer.

## 4.3 System software/communication

Communication with the acoustic sand detector system is on a RS422 serial line to the host computer. For the Norne installation, communication is via the subsea control module (SCM) to the process control and data acquisition system (PCDA) onboard the production vessel.

The communication with the sand detector system is a master-slave interface where the system responds to commands issued via the subsea control system (SCM).

Data is requested and transmitted to the topside PCDA system at regular intervals.
The following data is normally sent to the topside; the sand 'noise' level, the background flow 'noise' level, the rate of change for the sand and background flow 'noise' and alarm/status information.

Each time data is requested by the topside system, the parameters that set the frequency bands and threshold limit for the calculated sand 'noise' rate of change value are transferred to the sand detector system.

The system also includes a command that requests an averaged frequency spectrum consisting of 120 envelope values to be transmitted to the topside. The frequency spectrum covers the frequency range from 0 - 956kHz with a resolution of 8kHz.

The sand detector SW is stored in flash-eprom memory. Should it be required, a new program can be downloaded via the serial interface from the topside (PCDA). At Norne, the system is installed subsea on the 'Choke bridge', and can only be retrieved when this module is brought topside. Retrieving the 'choke bridge' is a costly operation and being able to carry out software changes by downloading new programs is therefore important.

## 4.4 Operator philosophy

The subsea sand detector system gives information about an increase or decrease in the concentration of sand particles carried by the flow, such that an operator can take appropriate action.

The average signal level in two frequency bands is calculated by the system. These signals are a function of the sand induced acoustic 'noise' and acoustic background 'noise' from the flow itself. The rate of change of these signals is also calculated by the system. The sand induced acoustic 'noise' is broadband and contains signal components within the frequency range of the sensor. The frequency range from 300 - 800kHz has been set as the default band for the 'sand 'noise' value. However, the band can be adjusted from the topside process control and data acquisition system.

The acoustic 'noise' induced by the flow itself is denoted 'background flow noise'. This 'noise' is expected to dominate at the lower end of the frequency range of the sensor, and the default range for this band has been set from 100 - 300kHz. However, the flow induced acoustic 'noise' is also influenced by the sand induced 'noise', but not to the same extent as the signals in the upper frequency range of the sensor.

The system as such, is not calibrated to provide accurate estimates of the current sand production over a range of flow conditions. Some form of calibration is possible after the system has been installed provided it is possible to obtain data about the amount of sand produced during stable flow conditions.

The calculated rate of change signal, based on the sand induced acoustic 'noise', is the main criterion for detecting a sudden increase in the sand production.

It is possible to develop models that can be incorporated in the sand detector system software such that quantitative measurements can be provided. However, the development of such models will involve a series of full scale trials using a test rig where it is possible to vary the flow conditions and to inject controlled rates of sand.

For a certain range of process conditions, models can be implemented in order to provide quantitative sand measurements. Such models can be implemented in software even after a system has been installed subsea, since the sand detector system has a facility for complete downloading of software from the topside.

Signal presentation

The signals representative of the sand induced acoustic 'noise' and the background flow 'noise' are displayed as historic trend curves. This is important as it gives the operator an indication of the relative changes in the sand induced acoustic 'noise' and the background flow 'noise'.

The rate of change of the signal for the sand induced 'noise' and the background flow 'noise' are calculated by the system. These values are displayed together with the historic trend curves for the sand 'noise' and background flow 'noise'.

Alarm criteria are based on the value of the rate of change of the sand induced acoustic 'noise' signal, but can also be based on the absolute level of the sand induced 'noise'. The latter value has to be set, based on operational experience as the sand induced 'noise' is influenced by the flow rate, flow regime and GOR.

An example of a historic trend curve for the sand induced 'noise' signal and the corresponding rate of change for this signal is shown in figure 5 below.

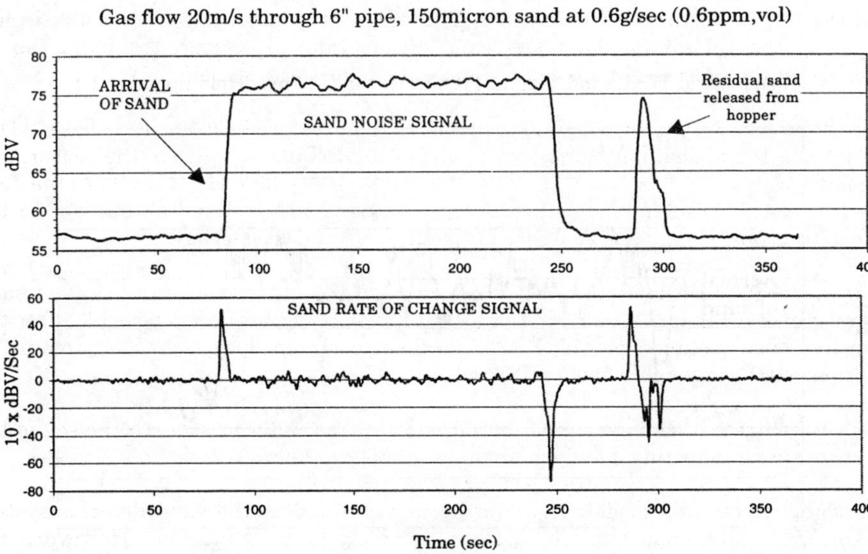


Fig 5 Example of sand induced 'noise' signal and sand rate of change signal

The example in figure 5 is from a test at AEA Technology in England and with the sensor attached downstream after a 90° bend in a 6" flowline. The example shown, is for a 20m/s air (only) flow and with 0.6ppm (vol) fraction of sand transported through the pipe. The curves have been generated by recording the broadband 'noise' signal from the sensor during the test, and later injecting the recorded 'noise' into the sand detector system. The broadband signal from the sensor was recorded during the test using a wideband instrumentation tape recorder.

Another example of a sand induced 'noise' signal is shown in figure 6. This curve is from a test at BP's Wytch Farm in England, where the flow from two oil wells was connected to a 2" test section. The sensor was attached downstream of a 90° bend in the flowline, and the signal from the broad band sensor was recorded using a wideband instrumentation tape recorder.

The recorded broadband signal has later been 'injected' into the front end of the sand detector system to generate the curve shown in figure 6. The curve shows the sand induced 'noise' signal for a test 'run' with 110micron diameter sand fed at a rate of 57.6kg/day into the 2" test section (about 25ppm fraction volume of sand). The average fluid flow was 17.63 $m^3$/hour and the gas flow rate at the temperature and pressure in the pipeline was 18.7$m^3$/hour. This corresponds to an average velocity of 4.7m/s for a 'hold-up' value of 1.1. The water cut during this test run was approximately 25%.

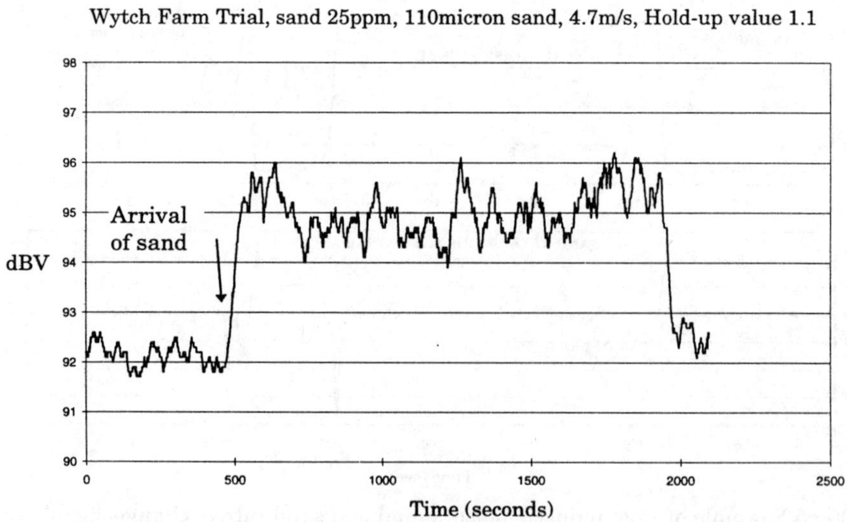


Fig 6 Example of sand induced 'noise' signal from multiphase flow

Information about the production flow can be displayed as a historic trend curve together with the sand 'noise' signal and the signal representative of the background flow 'noise'. It is expected that including information about the production flow rate, together with the signal representing the background flow 'noise', will be useful in that it may be possible to detect factors that impede the flow (e.g. the formation of hydrate plugs).

However, the main benefit of the system will be in continuously monitoring the production flow from a well during different flow conditions and to give a warning of an increase in the sand production rate.

**REFERENCES**

(1)Fossum T. G. 'Sand Detector for Subsea Installations', Subsea 89 Conference, London, Dec. 5-6, 1989, Organised by 'Themedia Ltd', (Sponsored by 'Subsea Engineering News')

# Underwater acoustic telemetry of complex analogue signals in a multipath channel

R S HABIB ISTEPANIAN
University of Portsmouth
B WOODWARD
Loughborough University

## SYNOPSIS

This paper presents a novel underwater acoustic telemetry system for transmitting multiplexed data in a multipath channel. Although any type of data, analogue or digital, may be transmitted, the system was designed principally for the transmission and accurate reproduction of analogue information. The communication link is achieved by a combination of pulse code modulation for data encoding and on-off keying for data transmission. Multi-channel telemetry is achieved by an inverse data rate interrupt multiplexing hierarchy in which a microcontroller in the transmitter forms a real time intelligent interrogator of each signal source linked to it. An identical microcontroller with the same interrupt priority software is implemented in the receiver, which is linked to a data logger for signal acquisition, display and further analysis.

## 1. INTRODUCTION

Acoustic telemetry of complex analogue signals representing behavioural, physiological or environmental parameters presents a challenging problem, especially to achieve high data rates in shallow water. To transmit and receive a truly continuous analogue signal in real time is impractical in most underwater environments because of the presence of a large number of possible paths, called multipaths, that the transmitted signals may take before reaching the receiver. This phenomenon is caused by reflection and scattering from the sea surface and the sea bed, and by volume scattering from marine life, objects, bubbles or inhomogeneities. The receiver can therefore detect a proliferation of unwanted signals, randomly time-varying in amplitude and phase, in addition to the wanted signal that arrives by the direct path from the transmitter. If there is relative motion between the transmitter and receiver, there will in addition be unwanted frequency shifts due to the Doppler effect. The situation is further complicated if two or more signals from different source sensors are transmitted.

Acoustic noise from shipping, surface waves, rain and marine life, and from thermal motion of the water molecules, usually limits the signal-to-noise ratio, which is rarely better than 20 dB in practice although this depends strongly on the transmitter-receiver configuration, sea depth, and so on. In principle, the signal-to-noise ratio may be improved by increasing the transmitter power, provided the cavitation threshold is not exceeded, but this exacerbates the multipath problem. Another method is to use directional transmitting and receiving transducers, or an adaptive receiving array that can track the transmitter.

These limitations can be overcome by choosing a suitable encoding format and a reliable detection scheme so that the transmission channel is effectively immune to multipath propagation (1-3). This means that a continuous signal has to be sampled before transmission but at a sufficiently high rate so that none of the information from the sensor is lost. For two or more signals sources, the transmitted signals have additionally to be multiplexed before transmission; this may be done by using a different carrier frequency for each parameter. In the system described here, both the transmitter and the receiver have a microcontroller to do the signal processing (2). They are linked asynchronously and each has compatible software and an identical priority protocol. The comparative simplicity of the hardware and software allows the system to be used in a variety of telemetry applications, such as transmission of subsea control data, diver monitoring and, with modifications, voice communications. Transmission of an electrocardiogram is presented as an example of an analogue signal that can be telemetered in a reverberant channel.

## 2. UNDERWATER CHANNEL CONSIDERATIONS

The two main considerations in designing a telemetry system are a multipath immune communication link and a reliable detection scheme. For any application the design is always a compromise. Usually the first considerations are the data rate and the range over which the information is to be transmitted. These factors determine the choice of the carrier frequency that may be used and in turn the physical size of the transmitting and receiving transducers (projector and hydrophone), their bandwidth, beamwidth and power capability. Other important considerations are the channel characteristics, which determine the signal-to-noise ratio and ultimately the signal level at the receiver, as calculated wth standard sonar calculations. The choice of frequency is crucial because in addition to the decrease in transmitted power from the projector by geometrical spreading, there is a further loss due to various absorption mechanisms. For example, at 50 kHz, a suitable frequency for many long range undersea applications (up to several kilometres), the absorption is about 12 dB/km, whereas at 500 kHz, suitable only for short range applications (up to several hundred metres) it is about 58 dB/km. The data rate for applications requiring the transmission of analogue information is usually very low. Many conceivable applications are concerned with status monitoring of signals that either remain fairly constant or change slowly (in electronic terms); for example, the output of a depth sensor used in the authors' work is sampled at 20 bit/s (2). This is a very desirable situation from a design point of view because the signals need only be sampled intermittently to allow accurate monitoring.

In most short range telemetry applications the direct path between transmitter and receiver is usually considered to be a straight line. In practice, the transmitted signals may travel through an arc or a combination of arcs depending on the sound variation with depth, the sound velocity profile. This does not usually matter as long as the receiver can still separate the direct signal from the accompanying multipath signals.

## 3. SYSTEM DESIGN

A complete system for the telemetry of data, digital or analogue, is shown in Fig. 1. As it is too complex to present in detail in a paper of this length, it will be explained only in outline here; the finer details of the system are described elsewhere (9). In this system, the electrical specifications are as follows: centre carrier frequency, 68.9 kHz; Q-factor, 4.1; bandwidth, 17 kHz; pulse width, 100 μs; data rate, 50 bit/s (variable by reprogramming); projector electrical power, 0.4 W (also variable); projector resistance at resonance, 276Ω; projector voltage 30 V(peak-peak); sensitivity, -201 dB re 1V/μPa; receiver voltage for 100 m range, 450 μV(peak-peak). The projector and hydrophone were both omnidirectional transducers, type D-170, Universal Sonar Ltd.

The analogue outputs of the various sensors and electrodes are amplified and each applied to a threshold and signal conditioning circuit that contains an analogue-to-digital converter. In the figure there are three channels of information but the design is easily extendible to more channels if required. The digitised signals are then fed to a microcontroller (87C51) that serves as a central processing unit whose multiplexed output is fed to a power transmitter and then to an omnidirectional projector. Three programmable oscillators, each with a different frequency, control the power transmitter. The three frequencies used here, 68.26, 70.217 and 72.28 kHz, are derived by binary division stages from a 4.9152 MHz programmable crystal oscillator; they are all within the -3 dB bandwidth of the projector, which is between 61.3 kHz and 78.3 kHz. They are effectively the 'label' for each transmitted parameter. The protocol of the transmitter microprocessor ensures that the parameter with the slowest rate of change has the highest interrupt priority; and that with the highest rate of change has the lowest interrupt priority.

The input to the receiver is a train of pulses from the hydrophone. These are amplified and fed to 8th-order band pass filters which separate them into three information channels determined by the carrier frequency of each 'labelled' parameter. The filtered signals are coupled to three time-controlled envelope peak detectors by the receiver microcontroller (also a 87C51) and then compared with adjustable threshold levels. The microcontroller, which has an identical priority protocol to that of the transmitter and is asynchronously linked to it, discriminates the detected direct path signals from the multipath signals and therefore decodes only those having a sufficiently large amplitude and with the correct encoded frequencies. In the system shown in Fig. 1 the captured signals are stored on a multi-channel data logger which can be linked to a computer for further analysis; this also allows a display of the source signals in their reconstructed analogue form.

There are many schemes for encoding digital data directly for underwater transmission. For example, amplitude shift keying (ASK), frequency shift keying (FSK), two-phase shift keying (2-PSK) and 4-phase shift keying (4-PSK). All have been used for data communications; and various forms of PSK have proved the best choice for high data rates, of the order of 10 kbit/s or more (4-6). In the system considered here the requirement is to transmit several sets of independent data simultaneously but at a comparatively slow rate. The main design problem is how to encode the data from each sensor so that it can be received unambiguously and with minimum errors. The approach used here is to encode all the data with one of the well-known 'analogue' techniques but to use a different carrier frequency for each parameter in order to 'label' them for ease of asynchronous decoding at the receiver. The following 'analogue' techniques, in which the amplitude of the transmitted pulses remains constant, are shown in Fig. 2:

*Pulse width modulation (PWM).* The time between the leading edges of the pulses is constant but the duration of the transmitted pulses is proportional to the instantaneous amplitude of the analogue signal. In a multi-path environment, data may be detected in error because of pulse lengthening.

*Pulse position modulation (PPM).* The duration of the pulses is constant but their position with respect to the sampling instant is proportional to the amplitude of the analogue signal. PPM is susceptible to the Doppler effect so is best suited to applications with a fixed transmitter and receiver. It is the simplest scheme to implement and has been used in some of the authors' work.

*Pulse code modulation (PCM).* Each sample of the analogue signal is converted into a multi-bit code, typically 8 bits, which represents the digital value of this sample. Although the scheme is more complex to implement it is more immune to errors from multipath transmission than PWM and PPM and was the preferred technique in the telemetry example described below (1-3, 7,8).

The signals transmitted through the water from sensors look like a random train of PCM-formatted pulses, each group having one of three carrier frequencies, as shown in Fig. 3. The technique is therefore a combination of time division multiplexing (TDM), pulse code modulation and frequency coding.

## 4. AN EXAMPLE OF TELEMETERING AN ANALOGUE SIGNAL

Examples of analogue signals that might be telemetered for subsea applications are variations of pressure from depth transducers or tide gauges, temperature from thermal sensors, gas partial pressure from a diving system, and physiological variables such as the electrocardiogram or respiration pattern from a diver. Telemetry of the electrocardiogram will be discussed here as an example of a complex, although repetitive, analogue signal.

The electrocardiogram is a train of signals that represents the electrical activity of the heart. Each of the transients, as shown in Fig. 4, is called the PQRST complex, which occurs every time the heart beats. This is a classic example of a signal that is only of use after being telemetered if it can be seen in its analogue form, as here. It is quite easy to transmit a single tone burst through the water for each heart beat; this would give a measure of the heart rate but would not give any other information which might be useful for diagnosis. To transmit information that would allow the PQRST complexes to be reconstructed it is necessary to sample the original signal at a suitable rate and transmit the samples. In the system described above, these are interleaved with signals from other sources.

The encoding process, which would be the same for any other type of analogue signal, involves PQRST detection and 'feature extraction', executed as a series of steps by the microcontroller. These steps comprise linear filtering, non-linear transformation, threshold detection, sample segmentation and feature extraction (3,8), as shown in Fig. 5. The program is initiated by reading a new sample of data every 5 ms, determined by the microcontroller's interrupt protocol. The 8-bit samples are fed to a 'moving window' buffer in the random access memory (RAM) of the processor. Since the interval of the PQRST complex is typically $<120$ ms, a moving window width of 100 sample blocks (500 ms) covers the whole complex. For every new sample of data (byte 1) fed to the memory, the oldest data point (byte 100) is removed and the new sample is passed through the detection process, i.e.the window advances by one sample interval. This process is repeated until the complete PQRST complex is within the time frame of the window. The program then freezes the current window and segments it into 10 samples which represent the main features of the original 100 samples. This is essential as it is not possible to encode and transmit all 100 samples in the period between two successive heart beats. The process minimises the required transmission time to a tenth of the original time needed, and therefore reduces the transmission sampling frequency by the same factor, i.e. from 200 Hz to 20 Hz.

Then the program moves to the transmission phase, which involves digital pulse code modulation (DPCM) and on-off keying (OOK). The stored digital samples are encoded into 7 bits and a block of 11 time slots of binary coded digits is mapped per transmission frame, including the code slot to provide word frame synchronisation. The initial 7-bit synchronisation (frame guard) code (111 1111) is to signal the receiver that the next ten slots represent new data. This is followed by serial transmission of the 7-bit coded data samples. The program first fetches the samples from RAM, then transmits the seven serial bits of data and outputs a 1 or 0 bit to the specified output port connected to the on-off keying transmission circuit.

The receiver is programmed with a complementary protocol to decode the received PCM data samples and reconstruct the electrocardiogram that was originally transmitted. The microcontroller acts as a digital decoder with two functions. First, the program identifies the synchronisaton (frame guard) code, then performs a serial-to-parallel conversion and queues the received samples into the ten original bytes of data representing the PQRST complex. Second, it controls the transfer of each sample block to a digital-to-analogue converter for display, monitoring and analysis. A system programmed as described here has been used successfully to telemeter the electrocardiogram of a diver in real time.

## 5. CONCLUSIONS

A system for telemetering continuous and intermittent data through an underwater channel has been presented. The principle depends on sampling an analogue signal and transmitting digitally encoded pulses using a suitable acoustic carrier frequency that depends on the range needed. The system has been designed for the transmission of multiplexed sets of data in a multipath reverberant environment and has been tested successfully. The precision of the reproduced analogue signals depends on the quantization used during encoding. Here, with 7-bit precision (128 levels), the signal resolution achievable is about 0.8% of the dynamic range, which is within the practical limits required for acceptable analogue signal reproduction in most scenarios. The encoding and transmission scheme described here represents a compromise but the data rate can be reprogrammed to suit the requirements of transmission and reception in channels with different multipath modes.

## REFERENCES


(1) WOODWARD, B. and HABIB R. S. Physiological monitoring by underwater biotelemetry, Proc. 1st European Conference on Underwater Acoustics, Luxembourg, 1992, 110-113 (European Commission, Edited by Weydert, M., Elsevier Applied Sciences).

(2) WOODWARD, B. and HABIB ISTEPANIAN, R. S. Acoustic biotelemetry of data from divers, Proc. IEEE 15th Annual International Conference on Engineering in Medicine and Biology, San Diego, 1993, 1000-1002.

(3) WOODWARD, B. and HABIB ISTEPANIAN, R. S. The use of underwater acoustic biotelemetry for monitoring the ECG of a swimming patient, Proc. 1st Regional Conference of the IEEE Engineering in Medicine and Biology Society, New Delhi, 1995, 4.107-4.108 (Biomedical Engineering Society of India, Edited by Guha, S.K.)

(4) BROCK, D.C., BATEMAN, S.C. and WOODWARD, B. Underwater acoustic transmission of low-rate digital data, Ultrasonics, 1986, 24, 183-188.

(5) CATIPOVIC, J.A. Performance limitation in underwater acoustic telemetry, IEEE J. Oceanic Engineering, 1990, 15, 205-216.

(6) FALAHATI, A., WOODWARD, B. and BATEMAN, S.C. Underwater channel models for 4800 b/s QPSK signals, IEEE J. Oceanic Engineering, 1991, 18, 12-20

(7) FRYER, T.B. A multichannel EEG telemetry utilising a PCM sub-carrier, Biotelemetry, 1974, 1, 202-206.

(8) HABIB ISTEPANIAN, R.S. Use of microcontrollers for diver monitoring by underwater acoustic biotelemetry in multipath environments, Ph.D. thesis, Loughborough University of Technology, 1994.

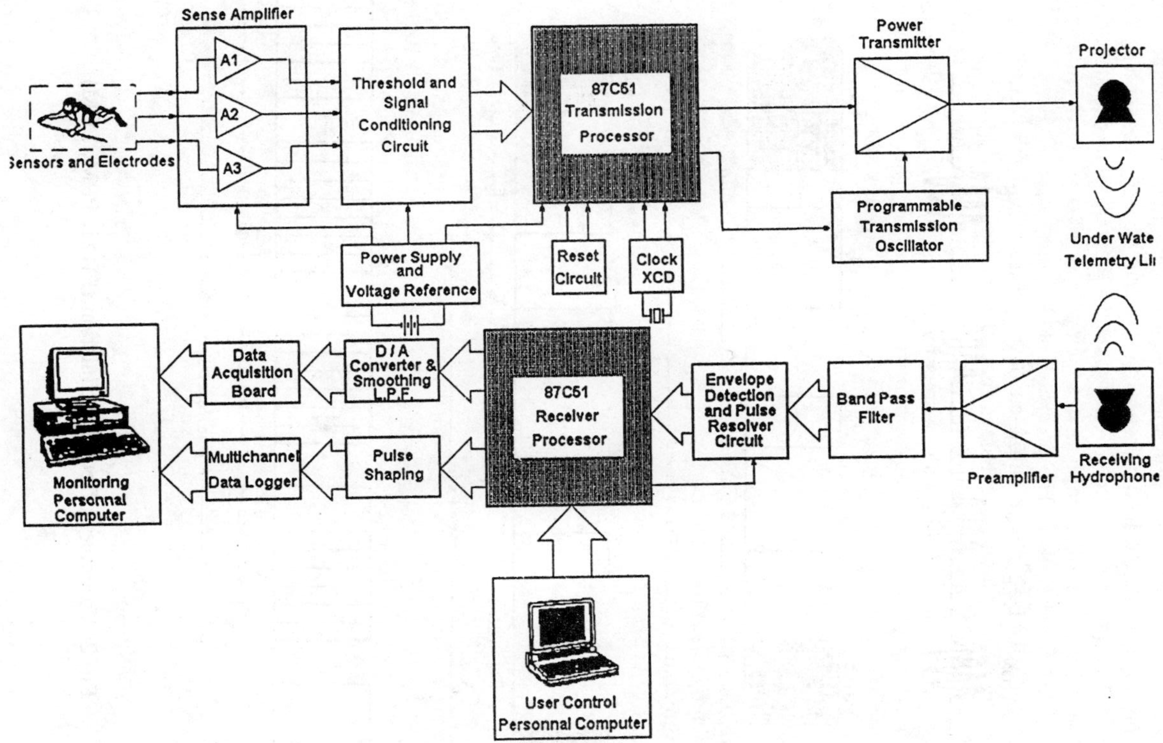


Fig. 1 Complete telemetry system

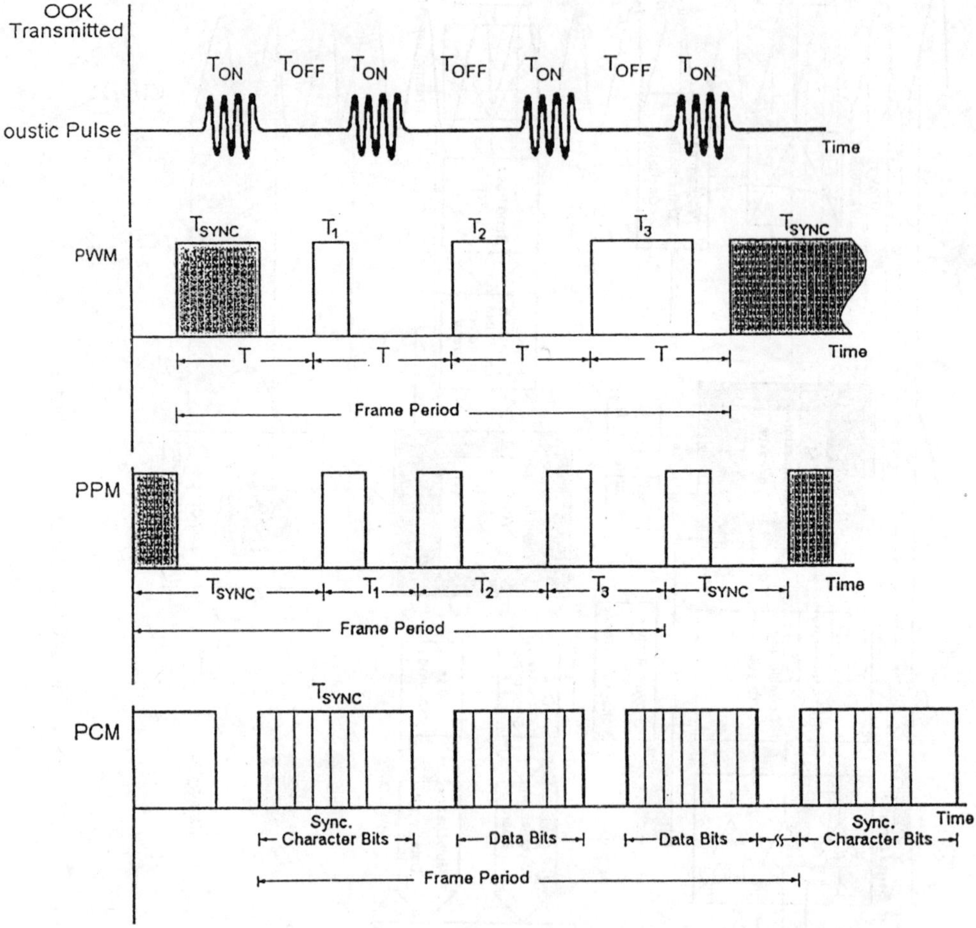


Fig.2 Examples of pulse modulation (PWM, PPM, PCM)

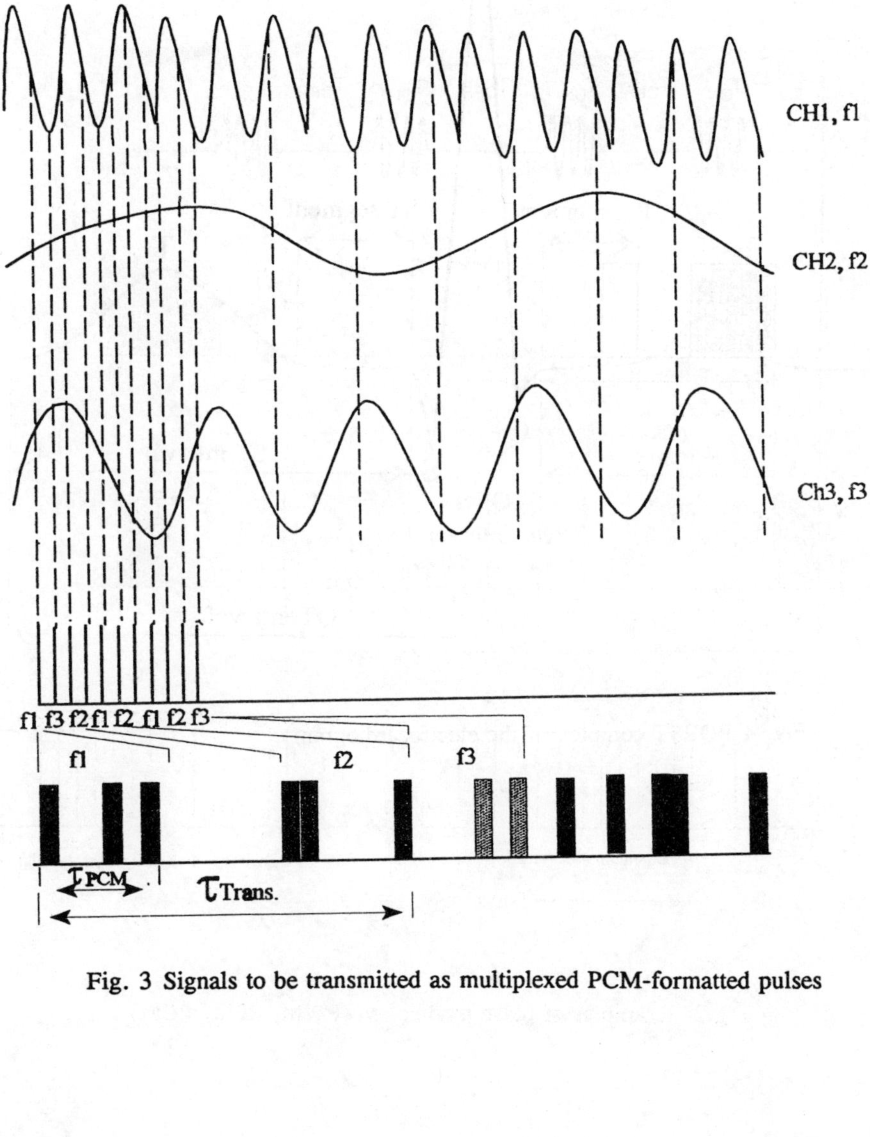


Fig. 3 Signals to be transmitted as multiplexed PCM-formatted pulses

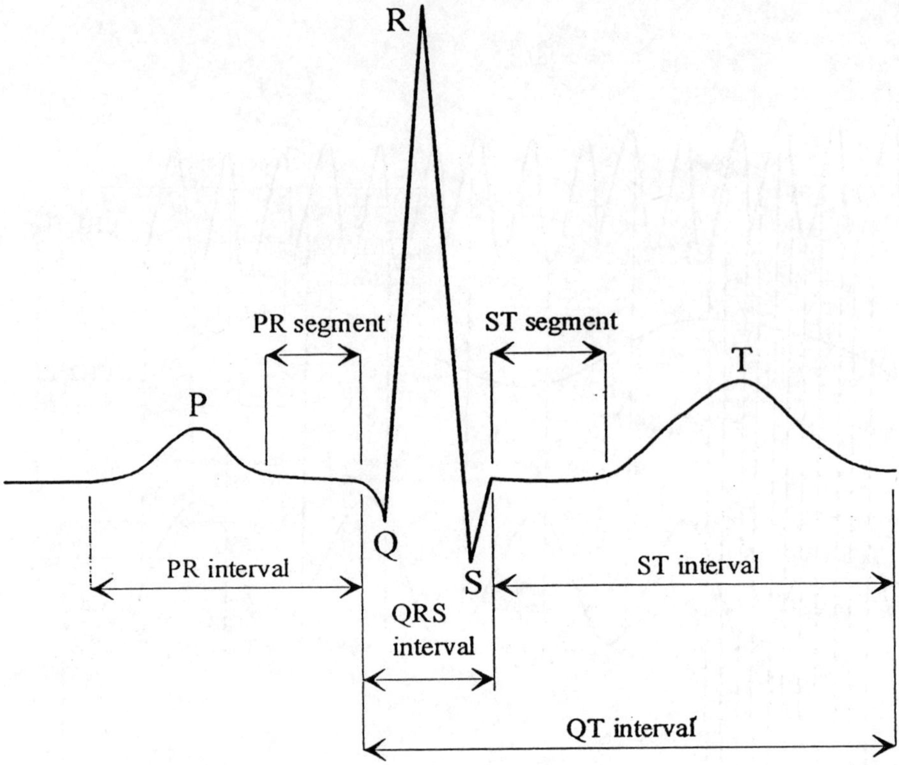


Fig. 4 PQRST complex of the electrocardiogram

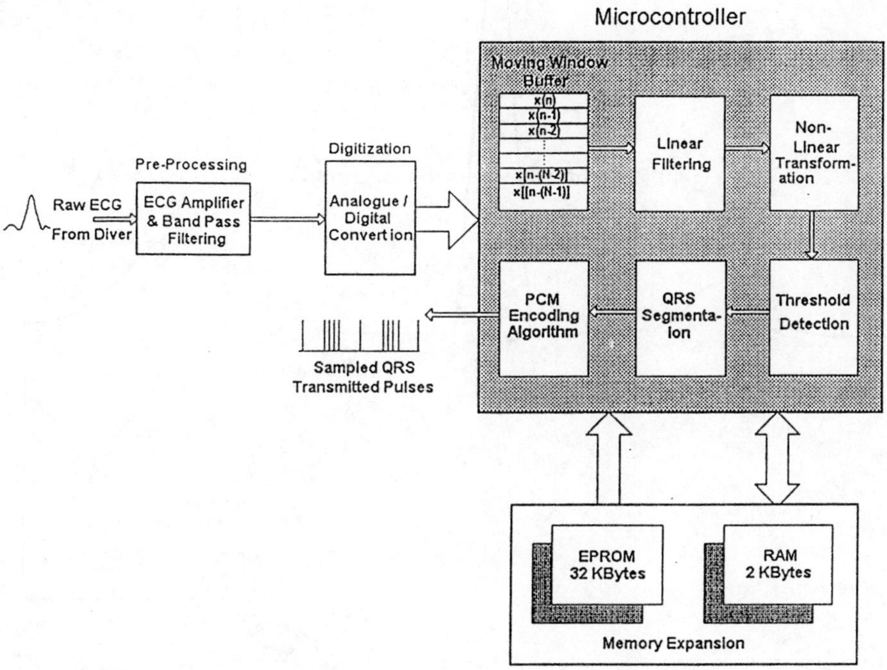


Fig. 5 PQRST detection and feature extraction process

# Damage detection in offshore platforms using vibration source location

M S GELDER and J M SHIPPEN
University of Birmingham

## SYNOPSIS

This paper investigates a vibration source location scheme for the global condition monitoring of offshore platforms. Vibration source location detects areas of potential structural damage through the change in vibration transmission path and the apparent change in the position of an active vibration source.

A platform is equipped at a number of locations with vibration sources and at different locations with monitoring transducers. A pattern recognition technique, minimised error functions, is developed using structural response vectors. This is compared to results using the Mahalanobis Distance and a Multi-layer Perceptron neural network approach.

The schemes are assessed using a 1/75th scale model of a North Sea offshore drilling platform.

## 1 INTRODUCTION

The condition monitoring of static structures and dynamic machinery has become common practice. Condition monitoring techniques offer the user the ability to track the state of a system parameter that is believed to be directly influenced by the system functionality, and many of these methods are concerned with vibration level and signature (either absolute or relative) as the system parameter. Benefits associated with condition monitoring include reduced production line disruption, planned machine down-time and speedier fault diagnosis.

Good progress has been made in the ability to predict rotating machinery faults such as shaft misalignment, gear and bearing wear and out-of-balance components (1, 2, 3, 4, 5). The condition monitoring of static structures has not been advanced by the same degree though. Even in the harsh conditions of Europe's North Sea the primary method of platform integrity monitoring is by the deployment of inspection divers. Using general visual inspection (major damage), radiography (cavities and inclusions), ultrasonic testing (structural member wall thickness and flooding) and magnetic particle inspection (surface discontinuities) this type of monitoring is labour intensive, location dedicated and hazardous. Furthermore, diver inspection is usually restricted to the so-called 'weather window', from April to September.

Investigations of deck based monitoring have been made using dedicated transducers (6), and mode shape and frequency shifts (7). The costs involved can prove prohibitive, and in addition, modal and analytic analyses are necessary. These are often difficult to interpret and correlate for complex structures.

The concept behind damage detection using vibration source location is as follows. A limited number of structural locations are monitored using transducers. The dynamic response from these locations is processed to form response vectors. Training response vectors are recorded when the structure, in a healthy state, is dynamically loaded at a number of locations. While remaining in this healthy state test vectors will correctly identify an active source through the association of the test vector with the appropriate cluster domain. Should the structure become damaged the vibration transmission path will be altered and the test vector will be associated with an incorrect excitation source.

An investigation into vibration source location has been reported (8) using a simple test rig. The scheme adopted a neural network pattern recognition approach, recording a 95% identification success rate.

This paper investigates a technique for source identification that is also based on pattern recognition principles. A scheme developed to enable multiple source activity to be identified is detailed (although only single source activity is considered within this paper) and is referred to as the minimised error function technique. In addition, pattern classification is performed using the Mahalanobis Distance measure (9) and a neural network approach (10). The test rig is a 1/75th scale model of Elf's DP2 gas drilling platform.

## 2 ANALYSIS METHODS

### 2.1 <u>Minimised Error Functions</u>

**In developing the method it is assumed that the structural response is a linear superposition of the deformations due to individual sources.**

**For a structure exhibiting training response vectors $v_i$ from $n$ active sources, an error will be generated when modelling a test vector $v_t$ with these training vectors. It is**

assumed that the test vector results through known source activity. The error can be expressed as,

$$\varepsilon = \left( \sum_{i=1}^{n} z_i v_i \right) - v_t \tag{1}$$

where $z_i$ is a scalar. In this form the error, ε, is a vector and is better expressed as the scalar

$$\varepsilon^2 = \varepsilon^T \varepsilon \quad \text{(for real vectors)} \tag{2a}$$

$$\varepsilon^2 = \varepsilon^T \bar{\varepsilon} \quad \text{(for complex vectors)} \tag{2b}$$

where $\bar{\varepsilon}$ is the complex conjugate of ε.

Values of $z_i$ are required that minimise $\varepsilon^2$. The differentiation technique depends on whether the function comprises real or complex scalars and vectors (if the vectors $v_i$ and $v_t$ are complex then the scalars, $z_i$, are also complex).

### 2.1.1 Real response vectors

The substitution of (1) in (2a) gives,

$$\varepsilon^2 = \left( \sum_{i=1}^{n} z_i v_i^T \right) \left( \sum_{i=1}^{n} z_i v_i \right) - \left( \sum_{i=1}^{n} z_i v_i^T \right) v_t - v_t^T \left( \sum_{i=1}^{n} z_i v_i \right) + v_t^T v_t \tag{3}$$

Differentiating with respect to the scalar, $z_i$, produces

$$\frac{\partial \varepsilon^2}{\partial z_i} = v_i^T \sum_{i=1}^{n} z_i v_i + \left( \sum_{i=1}^{n} z_i v_i^T \right) v_i - v_i^T v_t - v_t^T v_i = 0 \text{ (for min. error)} \tag{4}$$

which rearranges to give

$$v_j^T \sum_{i=1}^{n} z_i v_i = v_j^T v_t \quad \text{for } j = 1 \text{ to } n \tag{5}$$

For the purposes of this paper the real valued vectors represent the structural dynamic energy so (5) becomes

$$v_{kl}^T \sum_{i=1}^{n} \left( \sum_{j=1}^{n} z_{ij} v_{ij} \right) = v_{kl}^T v_t \quad \text{for } k,l = 1 \text{ to } n \tag{6}$$

This accounts for the energy being proportional to the sqaure of the frequency domain Fourier components.

### 2.1.2 Complex response vectors

For complex response vectors the substitution of (1) in (2b) gives

$$\varepsilon^2 = \left(\sum_{i=1}^{n} z_i \mathbf{v_i}^{\mathrm{T}}\right)\left(\sum_{i=1}^{n} \bar{z}_i \bar{\mathbf{v}}_i\right) + \mathbf{v_t}^{\mathrm{T}}\bar{\mathbf{v}}_t - \left(\sum_{i=1}^{n} z_i \mathbf{v_i}^{\mathrm{T}}\right)\bar{\mathbf{v}}_t - \mathbf{v_t}^{\mathrm{T}}\left(\sum_{i=1}^{n} \bar{z}_i \bar{\mathbf{v}}_i\right) \tag{7}$$

The expression is now minimised with respect to the real and imaginary components of $z_i$. In practice this is achieved by applying the Cauchy-Riemann conditions for the differentiation of complex functions.

$$\frac{\partial \varepsilon^2}{\partial Re(z_i)} = \frac{\partial \varepsilon^2}{\partial z_i} + \frac{\partial \varepsilon^2}{\partial \bar{z}_i} \qquad \frac{\partial \varepsilon^2}{\partial Im(z_i)} = \mathrm{i}\,.\left(\frac{\partial \varepsilon^2}{\partial z_i} - \frac{\partial \varepsilon^2}{\partial \bar{z}_i}\right) \tag{8}$$

$$\frac{\partial \varepsilon^2}{\partial z_i} = \mathbf{v_i}^{\mathrm{T}}.\sum_{i=1}^{n} \bar{z}_i \bar{\mathbf{v}}_i - \mathbf{v_i}^{\mathrm{T}}\bar{\mathbf{v}}_t = 0 \text{ (for min. error)} \tag{9}$$

$$\frac{\partial \varepsilon^2}{\partial \bar{z}_i} = \left(\sum_{i=1}^{n} z_i \mathbf{v_i}^{\mathrm{T}}\right)\bar{\mathbf{v}}_i - \mathbf{v_t}^{\mathrm{T}}\bar{\mathbf{v}}_i = 0 \text{ (for min. error)} \tag{10}$$

which becomes,

$$\bar{\mathbf{v}}_j^{\mathrm{T}} \sum_{i=1}^{n} z_i \mathbf{v_i} = \bar{\mathbf{v}}_j^{\mathrm{T}} \mathbf{v_t} \qquad \text{for } j = 1 \text{ to } n \tag{11}$$

It can be seen that expressions (5) and (11) are essentially the same, so the method of solution is the same for real or complex response vectors.

## 2.3 Mahalanobis Distance Measure

This is defined as,

$$D_{mah} = (\mathbf{v_t} - \mathbf{v_i})^{\mathrm{T}}\mathbf{C}^{-1}(\mathbf{v_t} - \mathbf{v_i}) \tag{12}$$

where $\mathbf{v_i}$ is the mean vector from a sample of $N$ datum $\mathbf{v_j}$ vectors, and $\mathbf{C}$ is the covariance matrix which is approximated by

$$C = \frac{1}{N}\sum_{j=1}^{N} v_j v_j^{\mathrm{T}} - v_i v_i^{\mathrm{T}} \tag{13}$$

### 2.3 Artificial Neural Network

There are a number of different types of Artificial Neural Network (ANN), one of which is the Multi-Layer Perceptron (MLP). This comprises a collection of simple processing units called neurons or nodes, that when combined enable more complex mathematical operations to be performed. Generally, the output from a node is found by the non-linear transformation of the nodal input. The logistic sigmoid function, producing values in the range 0 to 1, is such a non-linear transformation function. Figure 1 schematically depicts an MLP (2 input layer nodes, 4 hidden layer nodes and 3 output layer nodes).

An MLP is referred to as a taught network as for given input vectors the desired network response, or target vector, is known. Each nodal connection has a weight associated with it and the connection inputs are factored by this weight. All connection values inputing to a single node are summed prior to transformation to the nodal output. If the difference between the target and network output values is too great the weight values are adjusted, starting with the weights nearest the output layer. This is referred to as error back-propagation which was initially proposed by Rumelhart et al.(11).

## 3 TEST CONFIGURATION

There are three stages to the testing :

i. formation of source location class domains (training vectors)
ii. analysis of vibration spatial localisation (validation vectors)
iii. analysis of damage detection (test vectors)

Twenty-five samples were recorded for each excitation source location with the test rig in an undamaged state, these being the training vectors. Ten validation vectors were recorded for the intact structure and five test vectors for the case of structural damage. Ten vibration sources were considered.

Complex valued vectors considered the relative phase and magnitude components at the monitoring locations and were used for minimised error function analysis only. Real valued vectors considered the effective energy flow from source and were used for minimised error function, Mahalanobis distance and neural network analyses.

The test rig was a 1/75th scale model of the main support structure of Elf's DP2 gas drilling platform and is shown in Figure 2. The model comprised 8 main legs with 102 connecting members. It is manufactured from extruded square steel tubing and is fully welded. It is grounded on four rigid blocks via bolted brackets. A structural member was damaged by a through-the-thickness severance.

Ten excitation locations are considered and four monitoring locations. The structural response was measured using B&K type 4375 accelerometers interfaced with B&K type 2626 conditioning amplifiers. The response signals were low pass filtered prior to A/D conversion with a Kiethley Instruments DAS801 board.

The data acquisition program was coded in Visual Basic and analysis software in Fortran 77.

# 4 RESULTS & DISCUSSION

For this study a 4-10-10-10 neural network architecture was employed. Training used logistic sigmoidal transfer functions at internal nodes, linear functions at output nodes, and took approximately 20000 epochs to achieve a global network error of 1%.

The analysis results using the validation and test vectors (healthy and damaged cases) are summarised graphically in Figures 3 and 4 respectively.

It is evident from the results for the intact structure that the minimised error function technique with complex response vectors is the most reliable source location scheme. It is also clear, although not from all sources, that when damage is introduced it is clearly identifiable with the 95% identification figure dropping to 48%. It is significant that this reduction is realised principally by the total mis-identification of some sources with the others correctly localised which could lead to the identification of the damage location.

The real vector approach to error minimisation is not so successful. There are only three sources that provide indication of structural damage (2, 3 and 6) with source 6 behaving unpredictably (very poor intact identification; very good identification when damaged).

The Mahalanobis Distance measure and the MLP neural network approach both describe, to some extent, the characteristics of the cluster domain profile in *n*-dimensional space, where *n* is governed by the transducer coverage. The variation in their respective descriptions arise with the Mahalanobis Distance using statistical variance within the assumption that each class is described by a single domain, and the MLP generating an arbitrary number of non-linear decision boundaries. With the exception of source 6 the neural network approach is more reliable with the validation data.

It is worth noting that the neural network may indicate the presence of damage and still correctly identify the active source through a degradation in the networks ability to achieve target output values. In other words the taught neural network approach offers a goodness-of-fit measure.

## 5 FUTURE DEVELOPMENTS

It is concluded that it is feasible to detect the presence of damage in an arbitrary profiled structure, such as on offshore drilling platform, using vibration source localisation. The benefits of such a global condition monitoring technique are that limited monitoring equipment is required and the reduced need for analytical dynamic model development or correlation.

Although not presented within this paper it is noted from testing that on repairing the structural damage the source identification process is not necessarily returned to its original state.

Research is presently investigating methods for locating the damaged structural member. It is believed that intelligent methods such as fuzzy logic could prove effective. Furthermore, it is believed that the use of complex response vectors can be enhanced through the introduction of complex neural networks (12).

## REFERENCES


1. Gorter, J., Klein, A.J., Vibration Measurements Provide Condition Monitoring of Rotating Equipment, Oil and Gas Journal, 1986, Volume **84**(2), 64-70.
2. Angelo, M.R., Vibration Monitoring of Machines, Maintainance Management International, 1989, Volume 7(3), 135-154.
3. Dalpiaz, G., Meneghetti, U., Monitoring Fatigue Cracks in Gears, NDT & E International, 1991, Volume **24**(6), 303-306.
4. Tjong, J.S.Y., Lau, T.H.K., Chang D.K., Reif, Z.F., Monitor Engine Condition Via Vibration Signals, Proceedings-SPIE, The International Society for Optical Engineering, 1993, Volume 1923, 1252-1258.
5. Lowes, S., Condition Monitoring of Rotating Fans using Neural Networks, Proceedings of the First Annual Symposium of Postgraduate Research, School of Manufacturing & Mechanical Engineering, University of Birmingham, 1995.
6. Elf Petroleum Norge As. DP2 Lifetime Extension Study Report (537/SOV/EM).
7. Shahrivar, F., Bouwkamp, J.G., Damage Detection in Offshore Platforms using Vibration Information, Journal of Energy Resources Technology-Transactions of the ASME, 1986, Volume **108**(2), 97-106.
8. Shippen, J.M., A Technique for the Calculation of the Location of a Vibration Source, Proceedings of the Institution of Mechanical Engineers, 1993, Volume **207**(Part C), 273-278.
9. Tou, J.T., Gonzalez, R.C., Pattern Recognition Principles, 1974, (Addison-Wesley) 125-126.
10. Haykin, S, Neural Networks - A Comprehensive Foundation, 1994, (Macmillan College Publishing Company)
11. Rumelhart, D.E., Hinton, G.E., Williams, R.J., Parallel Distributed Processing Volume 1, 1986, (MIT Bradford Press) 318-362.
12. Masters, T, Signal and Image Processing with Neural Networks: A C++ Sourcebook, 1994, (John Wiley & Sons)

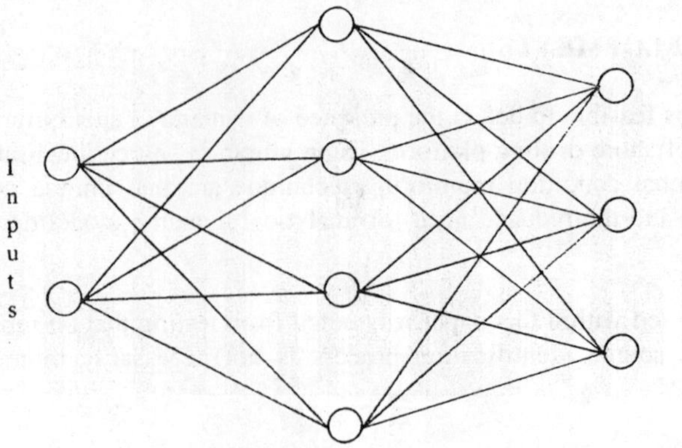


Figure 1 - Schematic of a 2-4-3 multi-layer perceptron network

Figure 2 - Offshore platform test rig model

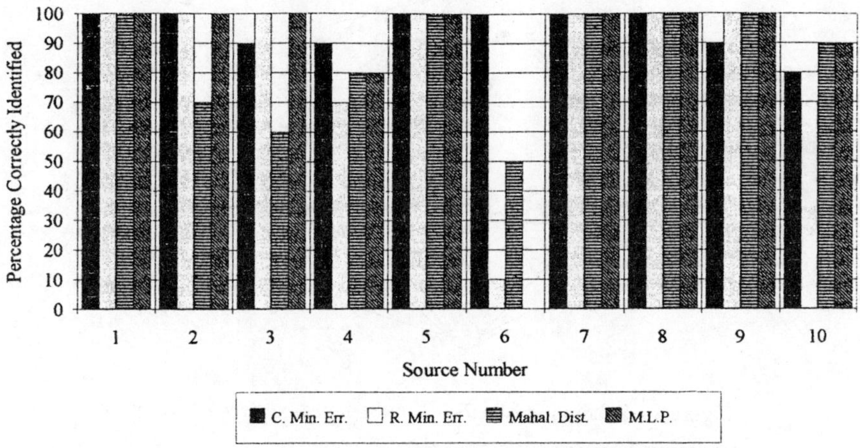


Figure 3 - Vibration Source Location Identification Results
Intact Structure
(C. Min. Err. : Complex vector minimised error functions, R. Min. Err. : Real vector minimised error functions, Mahal Dist : Mahalanobis distance measure, MLP : Multi-layer perceptron neural network)

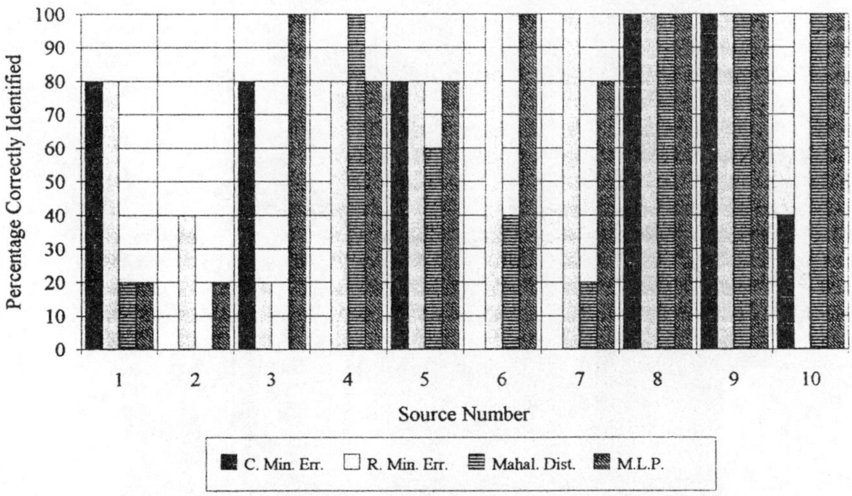


Figure 4 - Vibration Source Location Identification Results
Damaged Structure
(C. Min. Err. : Complex vector minimised error functions, R. Min. Err. : Real vector minimised error functions, Mahal Dist : Mahalanobis distance measure, MLP : Multi-layer perceptron neural network)

# .ow cost acoustic imaging using orthogonal echniques

M BERESFORD
EC-Marconi Limited, Templecombe, UK

## Abstract

Technology developments in military Sonar in the last 10 years have now evolved so that it is feasible to view underwater objects in turbid waters out to ranges of 50m with a picture quality which can be related to the photographic representation. It has enormous potential in black waters where underwater CCTV will not operate. Research based upon orthogonal imaging techniques and modern high frequency composite piezo-electric arrays will be described. The experimental system operated at near 1 MHz, provided 0.5° beamwidths with side lobes levels at better than -20 dB and produced processed images from underwater objects over the range 3m to 50 metres. The technology is currently available and suited for immediate development for low cost ROV application in offshore inspection, repair and maintenance. It is also suited for further miniaturisation for use by divers using head-up displays.

## 1. INTRODUCTION

The combination of recent efforts to provide affordable military systems, and the advent of low cost computing power in the form of PC technology, provides opportunities to exploit military investment in commercial and consumer applications. One such area is underwater viewing in turbid conditions.

Underwater imaging is usually achieved using optical means. However, in conditions of poor visibility, or even in clear water, where the range to the objects of interest is greater than a few meters, acoustics afford an alternative, provided that crucial differences are recognised as outlined in section 2.

Coastal, estuarial and sea where muddy sea beds are disturbed in particular, render underwater navigation and inspection difficult under poor seeing conditions. Even under extreme conditions, where optical range is reduced to below one metre, high frequency acoustic attenuation and scattering is low enough that an acoustic imager can form a useful adjunct if adequate image quality can be achieved. For physical reasons, high quality sonar imaging is generally more expensive than optics (for example CCTV). This paper describes a relatively low cost, high performance acoustic approach and results obtained with an experimental imager.

## 2. BASIC PHYSICS - THE OPTICAL ANALOGY

(i) The large wavelength differences cause the acoustic picture quality to be coarser than optics in practicable underwater imaging systems and the diffraction effects are more marked in acoustics. The diffraction effects of the aperture are, in angular terms, 1000 times larger in the acoustic case for a given aperture size. Thus, the sidelobes in an everyday optical image are negligible whilst in an acoustic imager they have to be kept under control if the nominal resolution is to be achieved. However even with suppressed sidelobes, in practice they tend to limit the dynamic range of an acoustic system: a bright specular reflection can, via sidelobes, destroy the image formed by other low level diffuse reflections. Another consequence of the acoustic wavelength is that more targets will appear "mirror like" to an acoustic wave than to light. Thus, very directional specular reflections are more likely to occur, and the desired image-forming diffuse reflections will have lower relative amplitudes than in the optical case.

(ii) True three dimensional information is available in acoustics, since range information is easily achieved by time of flight measurements and many materials are partially transparent to sound like X rays and thus it is possible to investigate the internal structure of some targets which are optically opaque.

(iii) The slow sound speed makes high frame rates difficult to achieve at low cost. This in turn makes eye-integration less useful and renders most sonars user-unfriendly.

(iv) Optical receivers are power sensitive square law devices, whereas practical acoustic sensors are linear devices; thus the latter retain phase information, enabling electronic beamforming techniques to be used and having the potential for true 3-dimensional imaging without resort to holography.

## 3. ACOUSTIC IMAGER OVERVIEW

### 3.1 Display formats

The three most commonly used are:

(i) 'B' scan (PPI) in which azimuthal and range information are collected and displayed.

(ii) 'C' scan in which azimuthal and elevation information at a given range is gathered to provide a 2 dimensional "photographic" image made up of pixels.

(iii) A 'volumetric' scan, in which both 'C' scan and range data are collected, giving the potential for true three dimensional imaging made up of voxels.

### 3.2 Imaging concepts (beamforming)

We may define three conceptual options for generating the pixels in an imager, which cover most of the practical imaging scenarios. We assume that the system is active; ie it has a transmitter of acoustic energy and a receiver (See Fig 1 and 2). The options have distinct characteristics in terms of imaging performance and costs. Table 1 illustrates this and compares the performance with a hybrid 'orthogonal' solution (Figure 2), sometimes referred to as a Mills Cross configuration.

The orthogonal solution has similar performance to Option 2 but requires only 2N transducers where $N^2$ is the number of pixels in a dimensional image (ie at a given range). The beamforming process involves only one-dimensional Fourier transforms in the hybrid system, as opposed to 2 dimensional transforms in the other options and also since there are only N transducers (not $N^2$), represents a substantial cost reduction.

Table 1 Comparison of imager options

| | OPTION 1 | OPTION 2 | ORTHOGONAL HYBRID |
|---|---|---|---|
| No of transmitters | $N^2$ | 1 | N |
| No of receivers | $N^2$ | $N^2$ | N |
| Frame Rate | Low | High | Medium |
| Directivity | Very High | High | High |
| 1st sidelobe levels (dB) (unshaded) | -36 dB | -18 dB | -13 to -26 dB |
| Cost | Very High | High | Low |

## 3.3 Beamforming

Acoustic imaging in the ocean at useful ranges beyond 10 m can only achieved at frequencies below about 2 MHz, where the wavelength is approximately 0.75 mm and is more than one thousandth times greater than in the optical case. This means that to achieve high resolution images of better than 0.5 degree, the acoustic aperture has to be several hundred mm. In addition, the high acoustic quality requirement implies:

(i) the use of shaded apertures to reduce the impact of sidelobes from specular reflections masking the desired diffuse returns which contribute to the quasi-optical image

(ii) detailed attention to array and electronic design to provide low inter-channel crosstalk.

Each pixel in the image requires the formation of a beam either on transmit or receive, or both, in the correct direction in object space. These beams can be generated electronically (eg, by temporal or spatial Fourier transforms or time domain algorithms; by mechanically scanning a single beam across the object space; or by any combination of these. Table 2 summarises the Options trade-offs in respect of cost, performance, size and complexity. The choice is between the flexibility and low wet-end volume of the electronic option and the low cost of the mirror system.

Table 2 Alternative beamforming/scanning technologies

| | ELECTRONIC | GEOMETRIC MIRROR | MECHANICAL |
|---|---|---|---|
| Cost | High | Medium | Low |
| Frame Rate | High | High | Low |
| Volume (wet end) | Low | High | Low |
| Reliability | - | Best | - |
| Complexity | High | Medium | Low |

Although at low frequencies, the bulk of mirror systems becomes prohibitive, they can compete with sophisticated phased systems above 100 kHz, and are the preferred low cost options at frequencies above 1 MHz. Past applications have included bathymetry (Haslett and Haliday (1), Knollman et al (2)), swimmer detection and minehunting. Further consideration of the trade-offs between mirrors and lenses indicate that mirrors have a potential performance superior to lenses and are easier to construct.

Simple cylindrical or elliptical mirrors can be used and the required geometrical arrangements can be achieved by offsetting the transmitter and receiver arrays relative to the mirrors.

### 3.4 Processing Concepts

A successful imager will operate in the continuous wave pulse echo mode, to minimise the effect of acoustic backscatter. The position of each volume pixel (voxel) in object space is identified with reference to the relevant transmitter and receiver array element numbers and the time from start of transmission.

The signal to noise excess for typical targets and ranges of interest has to be calculated by trading beamwidths, pulse lengths and frequency. The conceptual design assumes that each receiver array element samples a well defined solid angle in space and that range resolution is achieved by using short pulse transmissions and assumes good beampatterns are achieved by the use of well designed acoustic mirrors and transducer arrays which control sidelobes and (equally important) the cross coupling between array elements. Thus for the hybrid orthogonal approach (Figure 3) the total number of voxels is given by the product of the number of transmitter elements, the number of receiver elements and the number of range cells.

To minimise cost, low cost proprietary hardware and software is specified where possible. The use of a PC (or similar) with bespoke front end data acquisition and power amplifiers interfacing to the transducer arrays is a suitable starting point.

Figure 4 shows a conceptual view of the orthogonal receiver and transmitter multi element arrays interfacing with a PC for real time data acquisition. Figure 4 also shows the concept of transferring files from a real time data acquisition system to a more powerful work station for off-line analysis. The volume of data to be acquired depends upon (a) the frame rate (b) how the data is pre-processed (c) and visualisation/display methods.

(a) Frame Rate

The maximum frame rate for totally unambiguous imaging depends on the two way travel time to the range of interest plus a guard time to ensure that long range returns do not occur in the 'next ping' range bracket. Thus for option 2 (Figure 1) the maximum unambiguous frame rate will be of order 10's per sec, whilst for the orthogonal hybrid system this will be reduced by the number of pings required to insonify the desired field of view (in strips). Thus a reduction in complexity has been traded for a lower frame rate.

(b) Pre-processing Methods

Typical data processing options are illustrated in Figure 5. For example for profiling an object the 'first arrival detection' concept shown in 5(b) would be adequate, whereas a more complex data acquisition system for displaying true 3-D information to give quasi-optical quality will require a system similar to Figure 5(a), albeit at a cost in channel bandwidth. For the full-wave form presentation of data the word rate to the PC is dependant sample rate and could be of the order $10^4$ higher than the 'first arrival detection' method.(Ratio of sample frequency to frame frequency)

(c) Visualisation

The eye of the observer is adept at interpreting a poor quality picture when time integration of successive images is possible. This requires frame rates of several per second. Acoustic information is inherently 3-dimensional. This allows viewing of the scene from different points, either on a single display, or to generate true stereoscopic images.
Different forms of processing provide different interpretations of an object and in conjunction with 3D visualisations above, the operator can chose the best processing strategy to suit a particular imaging problem.

## 4 DESIGN BASIS

The principle of the orthogonal mirror beamformer is shown is Figure 3. The cylindrical mirrors act in a way analogous to optical mirrors and transducers are placed at the focal planes of both transmit and receive mirrors.

The experimental systems uses a pair of acoustic mirrors arranged as a Mills cross. One fan shaped beam 15°x 0.5° is transmitted and interrogated by the receiver array at the image plane of the orthogonal receiver mirror which has similar beamwidths. The image plane contains a composite piezo-electric transducer receiver array, the output of which is electronically multiplexed to sequentially sample and store the amplitude of the received signal. Subsequent pings are used to form a 2-D image. The 3-D image is then captured by either mechanical translating or panning. This multiplexing and panning arrangement was used to minimise the experimental cost, the object being to prove the acoustic principles. The frequency of operation was chosen to be 875 MHz to achieve the desired resolution out to 50m range, however, for shorter range applications the frequency may be increased to the benefit of underwater vehicle or divers sonar requirements.

The sonar head was mounted on a mechanical azimuthal training mechanism using a single transmitting transducer with a twenty four element array of receivers.

### 4.2 Reduction of sidelobes

The beampattern response to sidelobes was tailored. This can be achieved by adjusting the response of the transducers behind the mirror by adjusting the size and shape of the transducer elements. The theoretical on-axis beampattern response of an optimally designed combination of mirror and receiving transducer element is shown in Figure 6, and sidelobes of better than -25 dB have been achieved.

### 4.3 Receiver and Processing Display

The experimental system used a custom build pre-processor, commercial data acquisition cards and a PC AT with a mixture of custom and commercially available software for on-line processing and a Sun workstation running proprietary software for image visualisation and graphical rotation.

## 5 3D ACOUSTIC IMAGES OF FRACTURED PIPE

Preliminary experiments involved measurements of inwater-beam patterns and acoustic imaging of representation of a number of geometric shapes at ranges between 2.5 m and 25 m. These will be shown at the presentation.

One of the possible applications of the imager concepts concerns the inspection of underwater pipes in turbid (black water). According a target consisting of two sections of a 6 inch pipe separated by a 6 inch gap was imaged on a smooth sandy surface. Figure 7a shows a plan view of the 'fractured' pipe using a volumetric 3D visualisation technique. Although the line of sight of image was formed (taken) normal to the pipe, the visualisation software has allowed rotation of the image.

By rotating the same 3-D date field graphically different views may be displayed, Figure 7(b). Figure 7(b) illustrates a computer rotation of about 90° clearly showing the void in the pipe. Figure 7c shows an additional 90° rotation in about the vertical axis, now revealing a view inside the pipe down the axis. Finally figure 7(d) shows a further inspection inside the pipe from a viewpoint approximately 180° from that shown in Figure 7(b). The image Figure 7(a) was collected in non-real time and processed using a standard work station as described previously.

## 6 CONCLUSIONS

The experimental imager, the outline specification of which is shown in Table 3, has proved the concept of a low cost 3-D imager realised with a low hardware content and producing images with quasi optical visualisation of physical geometrics at ranges impossible with low cost optics. In particular images of a broken 6" pipe lying on the sea bed are shown indicating the X ray quality of the images for seeing through the object and being able to graphically rotate the image to see inside and round the back.

Further development using higher acoustic frequencies and commercially available processing should provide a low cost product suitable for civil and consumer applications including, for example, ROV inspection/maintenance platform installation within the oil and gas sector.

DTI/Industrial support for development is sought.

Table 3 Experimental imager outline specification

| | |
|---|---|
| Type of Imager | Orthogonal |
| Field of View | 12° (vertical) x 0.5° (Horizontal) |
| 4 dB Beamwidth | 0.5° |
| 1st sidelobes | Lower than -25 dB |
| Number of transmitters | 1 |
| Number of receivers | 24 |
| Beamformer | Acoustic Mirrors on Transmit and Receive |
| Object Ranges | 15-50m fixed focus<br>Adjustable down to 2 metres |
| Power Output | 1 watt |
| Operation Frequency | 875 kHz |

## 7 WHERE NEXT

The research concept has been proven for a system operating at 875 kHz designed for 50 m range. The same outboard system could be used with enhanced in-board data collection for real time operation (using mechanical pan for the transmit beam). Alternatively, a modified electronically scanned transmitter array can be implemented to effect a system without moving parts with frame rates at a few per second depending on range.

Table 4 shows the summary table from a short study showing variants of the experimental system with 12°, 24° and 48° fields of view operating at 2 MHz for short range applications appropriate to ROV and Diver Sonar.

Table 4 Specification for three field of view
(based on experimental equipment)

| | A | B | C |
|---|---|---|---|
| Aperture mm | 120 | 60 | 30 |
| Focus mm | 120 | 60 | 30 |
| Sidelobes | <-20 dB | <-20 dB | <-25 dB |
| Angular Field View | 12° | 24° | 48° |
| Angular Beamwidth | 0.5° | 1° | 2° |
| Depth of Field | | | |
| 5m focus | 3m - 10m | 2m - ∞ | |
| 3m focus | 2m - 4m | 1.6m - 18m | |
| 1m focus | 0.9m - 1.1m | 0.8m - 1.4m | 0.5m - ∞ |
| Approximate Dimensions (width x height x depth) | 200 x 150 x 200 | 140 x 80 x 140 | 160 x 80 x 80 |
| Autofocus | Yes for less than 3m | Yes for less than 1.5m | No |
| Operating Frequency | 2 MHz | 2 MHz | 2 MHz |
| Weight in Air (kg) | <10 | <10 | <10 |
| Range Resolution (mm) | 25 | 25 | 25 |
| Power Output (Watts) | 1 | 1 | 1 |
| Frame Rate | 1-3 Hz | 1-3 Hz | 1-3 Hz |

## 8 REFERENCES

Haslett RWG, Halliday W, "Rapid Hydrographic Surveying using Mirror Sonars". Proc. Conf. Electronic Eng. in Ocean. Technology pp 129-150

Knollman G C, Brown A E, Weaver J L, Bellin J L S, "Experimental Hydroacoustic Imaging System", J App.Phys, 42, 6, May 1971 p2168

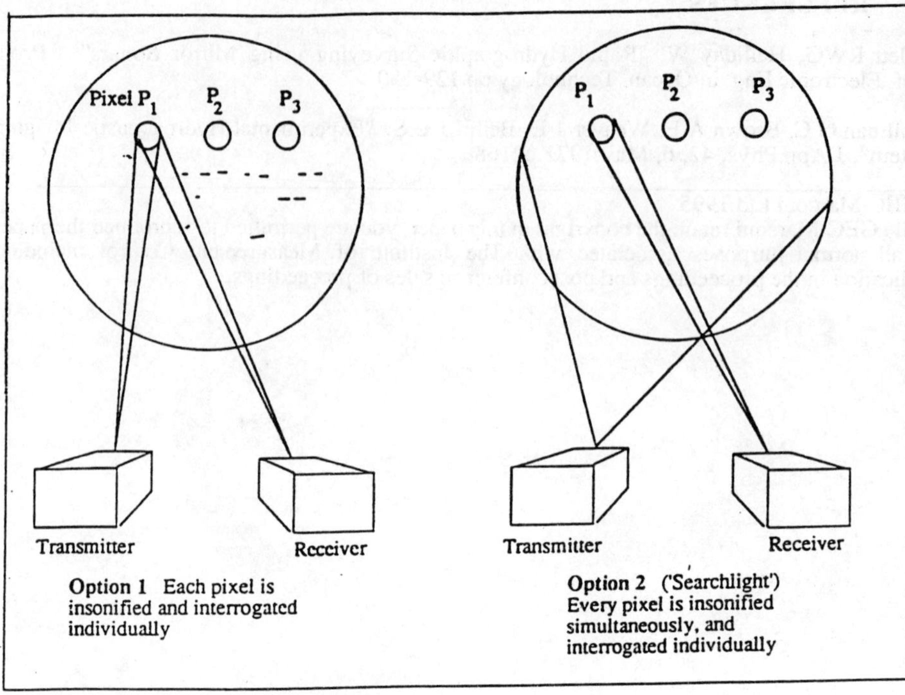


Fig 1 Imager Options

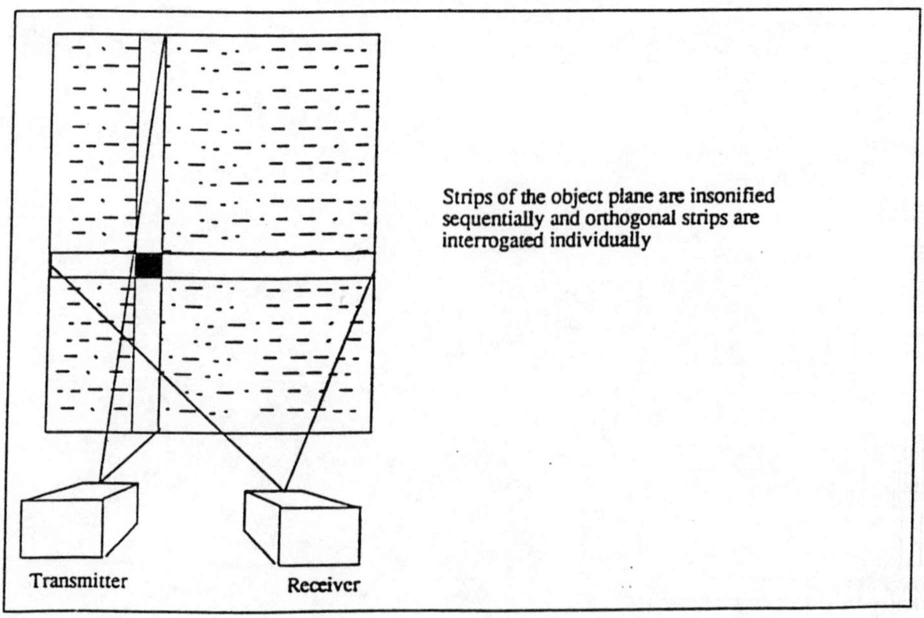


Fig 2 Orthogonal hybrid option

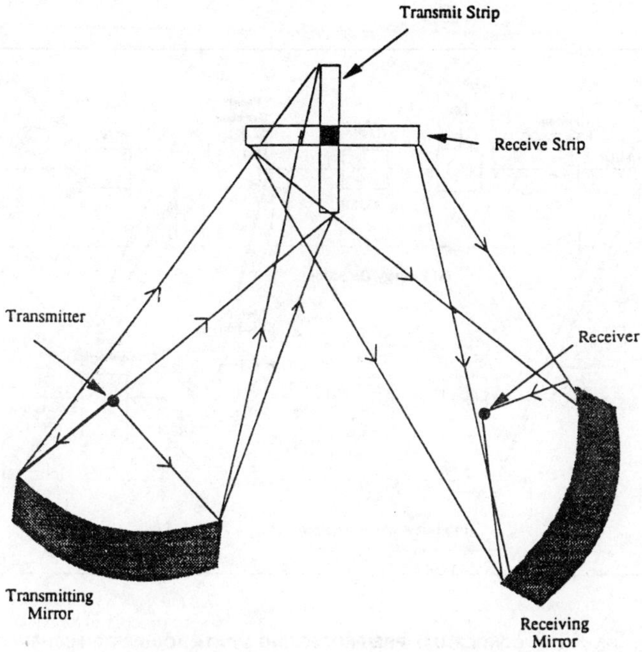


FIG 3 ORTHOGONAL MIRROR BEAMFORMERS

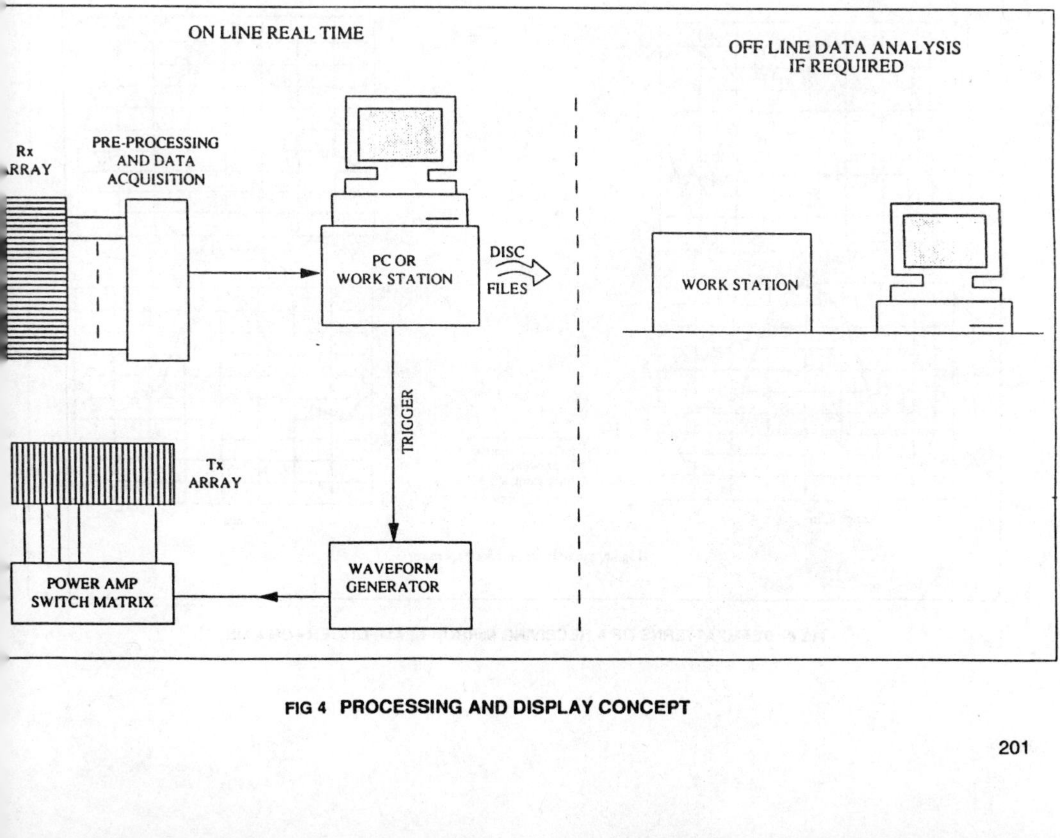


FIG 4 PROCESSING AND DISPLAY CONCEPT

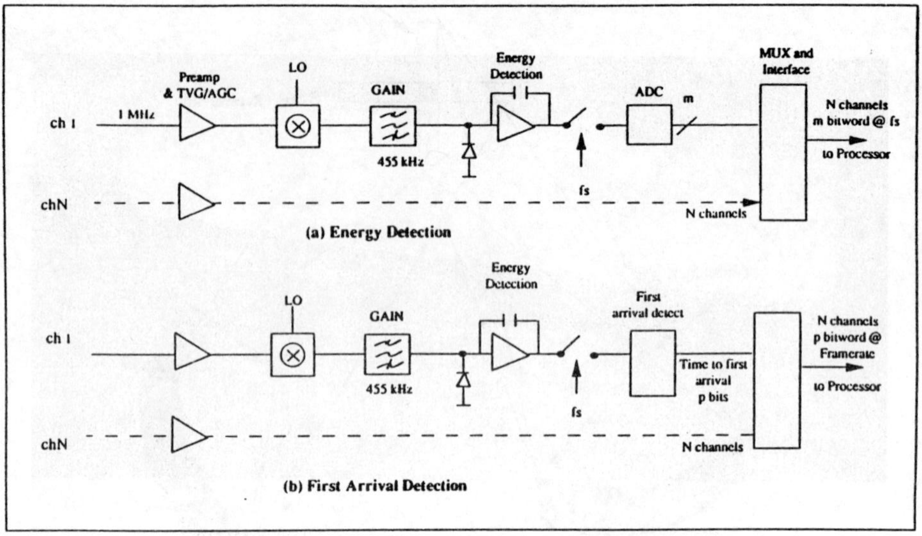


FIG 5 TWO CONCEPTUAL PRE-PROCESSING & DATA ACQUISITION SCHEMES

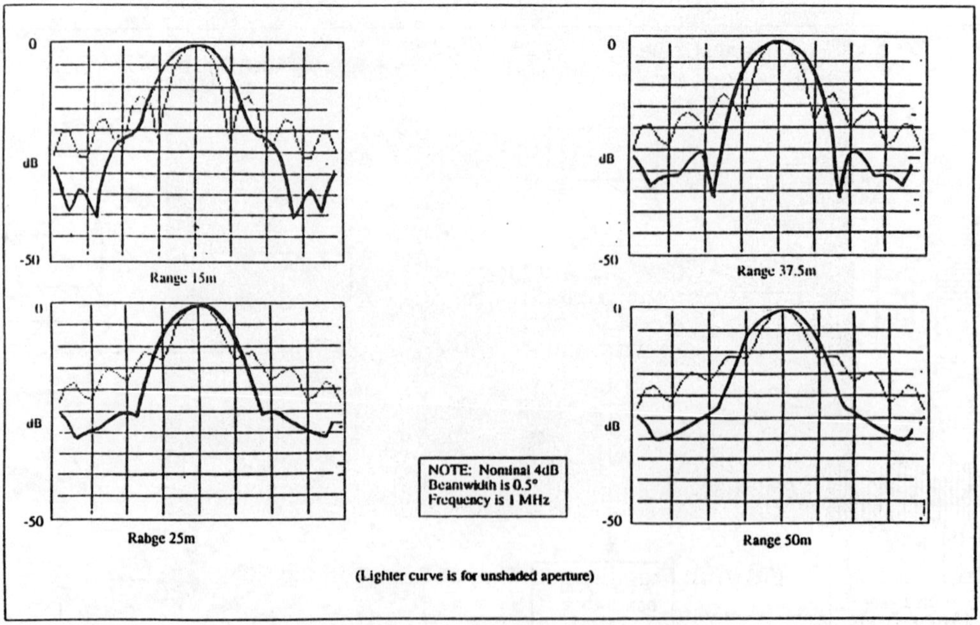


FIG 6 BEAMPATTERNS OF A RECEIVING MIRROR BEAMFORMER - ON AXIS

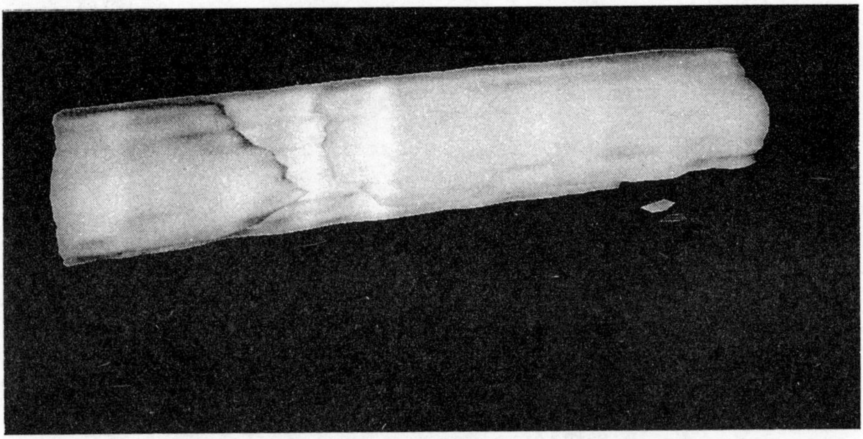

FIG 7(a): Fractured 6" Pipe at 3 m. Plan View

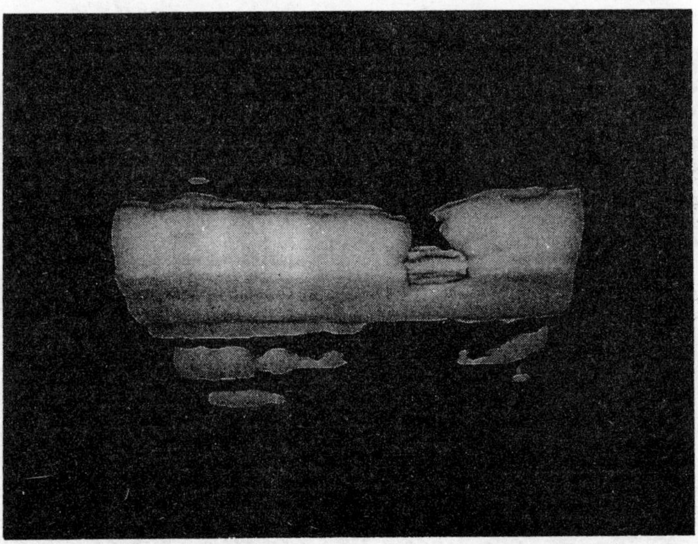

FIG 7(b): Fractured 6" Pipe at 3 m. Rotated 90°

FIG 7(c): Fractured 6" Pipe at 3 m. End View

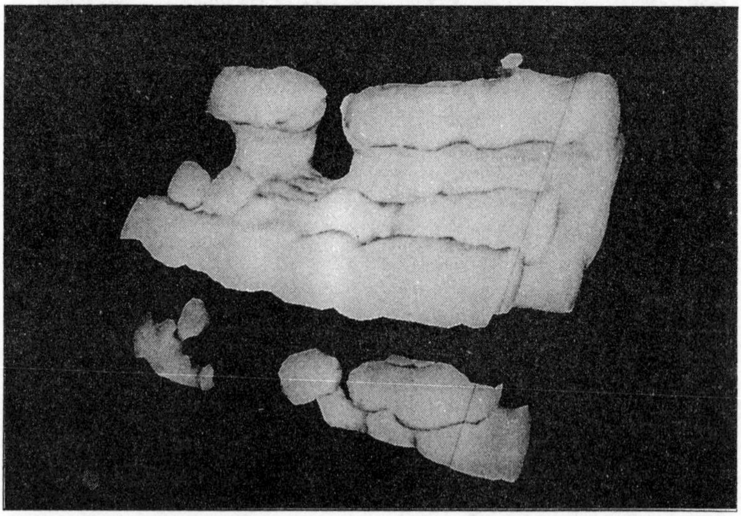

FIG 7(d): Fractured 6" Pipe at 3 m. Inside View

# High Integrity Pipeline Protection Systems

# ᵓipeline pressure specification break system

ᵛ D LOTH
ᵛ D Loth & Co Limited
BENJAMIN
BB Seatec

## ABSTRACT

This paper describes the configuration of a subsea High Integrity Pipeline Protection System, outlines the features which are important in the design of such a system, and presents a configuration meeting these requirements.

## INTRODUCTION

The concept is known by a variety of names. A High Integrity Pipeline Protection System (HIPPS) is also described as a High Integrity Protection System (HIPS), a Pipeline Pressure Specification Break System, and an Overpressure Protection System (OPPS). Regardless of its designation, the idea is not new; similar systems are used in the chemical industry, in petrochemical process facilities, and in surface production operations. The unique feature is the extension of the technology to subsea applications which creates a number of challenges. Origins of these systems lie in the requirement to control cascading reactions which develop into uncontainable processes.

Consideration of HIPPS for subsea application stems largely from the drive to develop high pressure high temperature fields using subsea technology. High pressure, high temperature, and corrosive fluids can mandate designs which executed in carbon steel are too thickwalled to be installed and, if executed in duplex materials require wall thickness beyond that currently considered feasible for welding. If the pipeline can be designed for a lower pressure, the development becomes feasible. In many other cases where conventional, full pressure rated, designs are readily achievable, substantial cost savings, perhaps in the range of £.5 M per kilometre, may be achieved. Further consideration of subsea HIPPS indicates that other applications are equally important. In many gas fields reservoir pressures drop very rapidly in the first few months, or year, of the development. The cost of pipelines designed for twenty year full containment pressure, at the initial value, is a frustrating expense. HIPPS although required only for the first, more highly pressured, months of development, could yield substantial savings. A further application is in conjunction with subsea separators. In the absence of venting capability found on the surface, extra care must be taken not to overpressurize the separator. A separator configured even for fairly normal production pressures would result in an extraordinarily heavy wall vessel which may be impractical, if not overly costly, for subsea application.

The underlying concept of subsea HIPPS is simple and straightforward. The process of reliably sensing and subsequently preventing further pressure buildup becomes far more complex when scrutinized.

Development of subsea HIPPS is complicated but many of the problems are amenable to detailed engineering. The electronic components are largely developed and require only adaptation for subsea application. Other components of the system may not currently exist but are probably extensions of designs suitable for lower pressure operations. Perhaps the most frustrating aspect of initial design work is lack of the information required to fully appraise the problem. There is an enduring scarcity of performance and reliability information for much of the required subsea equipment. This is not a new problem and one which is unlikely to be resolved in the near future. Ramifications of this will be more fully appreciated as discussion of system components proceeds.

## DESCRIPTION OF A PIPELINE SPECIFICATION BREAK SYSTEM

### General

A proposed Pressure Specification Break System, **Figure 1**, consists of three major subsystems; electronic control, a barrier device, and a "last chance" relief. Not all installations will necessarily require all elements and the design of these components could vary significantly depending upon application particular circumstances.

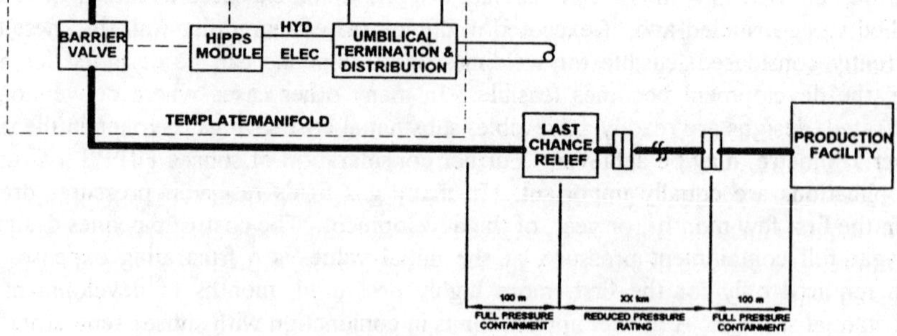


**FIGURE 1 BASIC PIPELINE SPECIFICATION BREAK SYSTEM**

Production equipment might be configured in several ways. The most common proposition is to collect production in a high pressure manifold and use a single HIPPS system to protect the pipeline. While this approach is superior on a projected reliability basis, due to the fewer number of safety critical components, it may have the significant disadvantage of relying on more unproven high pressure components. An alternative is to protect individual wells. By doing this some theoretical loss of reliability is unavoidable but the wells can be equipped with currently available equipment. Selection will have a cost impact, but a more significant effect will be the manner in which a "safety" zone of full pressure containment is maintained around the wells.

A wide range of possibilities for a Specification Break System, such as parallel vent lines or venting into adjacent wells, has been debated for some time[1] but most current consideration is given to a system similar to that described above.

**Electronics**

The underlying principles of sensing the pressure buildup and isolating in the pressure source are well understood. Fundamentals of the triplicated, voting, systems seem largely agreed. One of the major problems to be addressed from the electronic aspect is the inaccessibility of the subsea location with attendant restriction on maintenance. Designs housing the primary control electronics on the surface have been considered but these carry the substantial risk that the communication link between surface and subsea installations could be disrupted rendering the safety system inoperative. There now seems to be general agreement that a dedicated safety system is required and that the heart of this must be subsea; debate continues on how this can be best achieved.[2,3,4,5,6,7] It does not appear that the system must be entirely self-contained but certainly the primary functions of sensing the pressure buildup and activating the device to prevent further pressure increase must be a standalone subsystem.

Under normal circumstances the dedicated safety system is not required when the production control system is inactive. The significant exception is when production is supposedly shut-in and pipeline/pressure buildup occurs. In this situation, the system needs only to activate itself and operate the safety device; routine communication, testing, and restart operations need only be conducted when the production system is operative or at least capable of operation. This means that the bulk of subsea and surface communication can be conducted through the existing production control system; only the functions of sensing pressure buildup and actuating the safety device need be autonomous.

The control aspect, incorporating both electronics and hydraulics, is shown in **Figure 2.** One feature which will be noted is triplicated pressure sensors. Pressure sensors do not have an enviable track record subsea and every step must be taken to avoid common mode failures;. these are unlikely to be faulty sensing elements and more likely to be related to the masking of pressure buildup due to blockage by sand, paraffin, or hydrates. Careful placement of the sensors will be required.

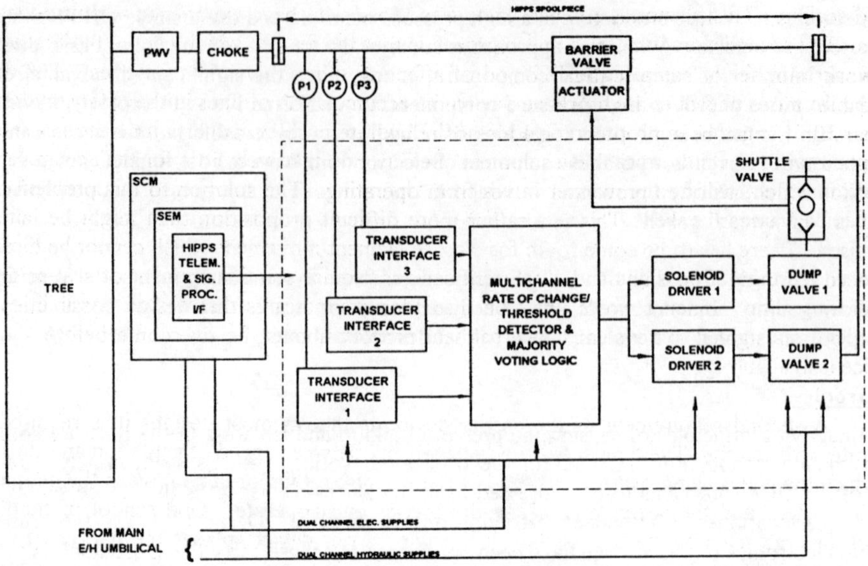


FIGURE 2 HIPPS CONTROL & MONITORING SYSTEM

**Barrier Device**

The functional specification of the barrier device remains unsettled; there may be unanimous agreement that flow must be shut off but how much and how well remains unsettled. The results of some analyses have indicated that it is small amounts of leakage which constitute the major threat to the pipeline. If this be true, the most important feature of the pressure barrier must be a positive shutoff. The actual requirement varies materially with the fluid being considered. While the leakage through any given defect might be greater with gas, under most circumstances a very substantial amount of leakage for some duration of time will be needed to compress the gas in the pipeline to a dangerous level. A much smaller leakage in a black oil system could lead to rapid overpressurization of the line. If the problem is not perceived as one of leakage, but rather a rare common mode failure in a production control system which allows one or more wells to flow, the primary requirement of the device is an extraordinarily high probability of closure rather than an extraordinarily low leak rate.

Regardless of the requirement posed, the shutoff valve, or barrier device, must have some unique capabilities. The concept of simply placing additional valves in series may appear attractive from a numerical sense but it is almost certainly inferior in some respects. In the first case there is little likelihood of serial valves providing greater sealing capability than the two valves, wing and master, which are currently incorporated in the trees. This is because, with the exception of a wedged gate valve, the partial closure or full closure with substantial leakage of the first valve in the series, normally removes the bulk of differential pressure loading required to shift the components in either a slab gate valve or a ball valve to effect good sealing. This phenomena is well known to those who have experience with production trees. The second drawback with this approach is that the feature which causes the first valve to work imperfectly has a high likelihood of affecting other valves in the series. The most probable causes of failure are blockage due to the accumulation of fines in the cavity, hydrates, or paraffin. It has been proposed to address the hydrate problem either with trace heating or circulation of an inhibitor but these solutions obviously do relatively little for the accumulation of solids which has often prevented valves from operating. The solution to this problem may lie in a "cavityless" valve. This is a rather more difficult proposition than might be initially surmised. There has to be some room for closure element movement which cannot be directly exposed to production fluids and all current designs require some movement of seal surfaces to allow sealing. Initial concept work on these designs indicates that design possibilities do exist, but substantial material and external seal problems, must be overcome before such a device is available.

A second requirement of the barrier device is that in most designs it is required to operate with alacrity. There must be some relationship between speed of actuation and the rate at which the line is being overpressurized but there is also a relationship between the geometry of the system and the speed with which the barrier must operate. One school of thought, which favours manifolding the wells with a single barrier device, places the barrier valve at some distance from the manifold to allow sufficient time for closure. If individual wells are to be protected, and flowlines are of only modest length, the valve will have to operate substantially more rapidly. Designing an actuator to stroke the valve in a few seconds need not be a difficult proposition but the valve itself will require some buffering mechanism to counteract inertial effects.

**Pressure Relief**

Although called pressure relief, the problem of ensuring that any failures do not exist in an area adjacent to the well or platform in which personnel might be endangered could be addressed in several ways. The probability of using either the relief, or other designated failure, should be very small. Many designers have addressed this problem by simply adding a length of pipe, rated to full pressure capability, for a length extending beyond the perimeter of the zone in which full pressure containment is required. While this addresses the safety problem quite adequately, and eliminates any failure modes associated with a mechanical relief, it does have some drawbacks. The first of these is that if there is a failure, its location is unknown. It could be located anywhere in the lower pressure rated segment of the line. The second problem is that if the line does rupture, it will continue to void to a point well below the pipeline's design pressure. A properly designed mechanical relief should contain any discharge from the line to the minimum amount required to maintain the integrity of the rest of the line. Besides the obvious benefit of minimizing emissions, this may have the significant advantage of avoiding the requirement to dewater the line. A second advantage of the mechanical relief is that any subsea intervention is at a predetermined location and can thus be more readily subjected to contingency planning than a random pipeline rupture.

## SUMMARY AND CONCLUSIONS

There is a substantial amount of work to be completed before HIPPS systems are widely utilized subsea. The prize, pipeline cost savings, may be high but so are the attendant risks. Few operators would wish to be accused of undertaking untoward risks, threatening the environment, for the sake of even large savings in the pipeline. Less risky, shorter term, applications on gas lines which will only be over pressurized for a brief period, may be required to build confidence in the concept. There are a number of aspects, from an operational vantage point, which will require careful thought. Shutdown of the system is the most obvious requirement and receives the most attention but restart of production probably poses the most challenging problem. In the first place, downhole safety valves will have to be repressurized. Since the pipeline itself would not be capable of withstanding these pressures, an alternative would have to be found. One option, often quoted, is that in any event the high pressure chemical line will be required. This line could be used to repressurize, that is equalize, the downhole safety valve prior to opening. The second aspect, which is yet to receive substantial attention is the overall question of availability. A safety system, which by necessity must be very sensitive, may also prove susceptible to a number of spurious shutdowns. Each of these shutdowns has a very finite cost in terms of lost production and the expected value of these shutdowns could rapidly eclipse the cost savings achieved by the pipeline pressure spec break system.

A number of study efforts are either currently underway or proposed to address the problems remaining in subsea application of HIPPS systems. It will probably be necessary to demonstrate these systems in conjunction with other cost saving design improvement, such as limit state pipeline design, before they will be interested as the sole mechanism for overpressure protection. Until that phase has been completed, the full cost reduction potential may not be realized.

## REFERENCES


1. *"Study of Pressure Specification Break in Subsea Production Systems"*, Pell Frischmann Engineering Ltd. Project 7081 DTI/OSO UK.
2. Bayly, D B and Loth, W D: *"Pressure Specification Break Pipelines"*, Society for Underwater Technology Conference Aspect '94 - *Advances in Subsea Pipeline Engineering and Technology*, Aberdeen Scotland 12-13 May 1994.
3. *"Subsea OPPS Feasibility Study"*, SINTEF, Norway 1993-94.
4. Onshus, T., Aaro, R., and Lund, B. F: *"HIPPS Applications and Acceptance Criteria"* OTC Paper 7828, OTC Conference Houston 1- 4 May 1995.
5. Lund, B. F., Onshus T., and Aaro, R: *"HIPPS Concepts for a Subsea Field Scenario"*, OTC Paper 7830, OTC Conference Houston 1- 4 May 1995.
6. Aaro, R., Lund, B.F., and Onshus, T: *"Subsea HIPPS Design Procedure"*, OTC Paper 7829, OTC Conference Houston 1- 4 May 1995.
7. High, G. and Loth, W. D: *"A Complete High Integrity Overpressure Pipeline Protection System"*, OTC Paper 7832, OTC Conference Houston 1- 4 May 1995.

# igh integrity solid state shutdown systems in ubsea locations

RAMSAY-CONNELL
'I Industrial Automation Limited

## 1 Introduction

The paper will describe a High Integrity Solid State shutdown system incorporating partial and full closure test facilities for location subsea. The issues of relevant safety standards and the means of demonstrating compliance will be summarised, with a description of the actual application of Inherently Fail Safe solid state logic being located subsea and controlled from the surface via communications.

The issues of isolation of safety functions from control and monitoring will be described, along with identifying the manner in which test functions do not inhibit genuine trip demands.

Facilities for the testing of inputs, the management of faulty transmitters and the instigation of partial closure testing on demand are included.

In built functions for High Resolution Sequence Of Event logging for post incident analysis are also described.

## 2 The Issues

The current economic climate is resulting in a number of initiatives intended reduce costs of crude oil production, which if successful will lead to development of more marginal fields. Two techniques currently being aggressively pursued are the use of Floating Production, Storage and Offloading (FPSO) combined with subsea completions. The latter item is the focus of this paper, but the requirement for subsea Emergency Shutdown facilities is a late comer to the scenario.

It may be questioned why there is a need to even consider subsea HIPS, and it is apparent that the attributes of High Integrity Protection Systems (HIPS) for subsea applications were perhaps not initially addressed to great detail, after all, there is little risk to human life in the immediate vicinity, with the shutdown logic (if any) tending to be located topsides. A number of factors now influence the need for HIPS on subsea systems, some technical, some ethical, some political and some commercial. They can be summarised as follows:

- Pressure To Reduce Production Costs.
- Use Of FPSO Methods Combined With Subsea Completions.
- Pressure To Improve Operational Safety.
- Pressure To Provide Auditable Safety Functions.
- To Reduce Flowline Costs By Preventing Over Pressurisation

It is certainly the latter item which has had the greatest influence on the current interest, but the ability to simultaneously address the other factors is being recognised. Figures for the cost of flowlines designed to withstand overpressures can run into tens of millions of pounds per kilometre, whereas an all inclusive high integrity HIPS system is costed in the tens of thousands of pounds.

The drawback of the current usage of shutdown logic located topsides is the dependence on the communications via the umbilical. Hardwired connections can be made, but the increased cost of the umbilical raises questions over the practicality of this.

One alternative is to use the subsea control package to perform the shutdown functions, thereby utilising the communication link for the relay of information, and this has been used in a number of applications. The advantages of this are clearly the savings in space, hardware and power.

This combining of safety and control functions, however, is not an approach accepted easily in normal operations topsides, and the same arguments are being brought to bear on the subsea designs. Particular concerns relate to the lack of diversity, the inability to quantitatively assess the safety level, common mode faults and the ability to generate a sound safety case.

The onus to add another system to the subsea package creates its own problems in terms of logistics and utilities. Any system must meet the constraints of power loading, weight and size as well as being cost effective, and above all, demonstrably safe.

These issues combine to define the constraints for a subsea HIPS system which are as follows:

- High Integrity
- High Availability
- Redundant Inputs
- Diverse From Control Functions
- Remote Communications
- Autonomous Safety Functions
- Remote Testability
- Space
- Weight
- Power
- Cost

Solid state technology is often mistakenly seen as being "old fashioned" and inflexible. In this case however, it is demonstrably the optimum solution to these issues.

# 3 The Solutions

Taking each of the above issues in turn, the purpose of the item can be defined and the manner in which it is achieved described.

## 3.1 High Integrity

The definition of integrity is the confidence that a system will do what you want, when you want, every time, under a wide range of conditions (and conversely that it won't do what you don't want any other time). Substantiating integrity in a quantitative manner is an ongoing dilemma for any safety systems including software, as only qualitative arguments can be presented if the programming has any level of complexity. In addition the effectiveness of the on-board diagnostics is not 100%.

For solid state systems, however, there are well documented and auditable means of calculating the failure rates of the components based on widespread usage of those components in thousands of applications.

The most commonly used is Parts Count Analysis (PCA) based on MIL.HDBK.217, which many will argue is not ideal, but has the advantage of being easily understood in principle by non-academic persons. Further, when applied to safety systems using full Parts Stress Analysis (PSA), the results tend to be pessimistic as they do not take account of circuit design. The actual tolerances permitted by components in safety systems is often much wider than that allowed in the MIL.HDBK.217 calculations.

The PCA/PSA methods do not differentiate between types of failures, which for safety systems must be split into Nuisance Failures and Unsafe Failures. Nuisance failures will be covered later. Unsafe failures are those which are unrevealed by the system diagnostics in normal operation and result in a failure to operate on demand when a hazard is detected.

In a software based system the number of failures detectable by the diagnostics is defined by the Coverage Factor 'C', a figure which is rarely if ever provided by the system manufacturer as there is little agreement on the best tool to define it.

In High Integrity solid state systems however, the detection is not based on external diagnostics, but is inherent in the system normal operation. This means that one key part of the auditability dilemma is effectively taken out of the picture.

Knowing the coverage factor is all very well in theory in terms of estimating the unsafe failures, but what yardstick should be used to define the acceptable level of unsafe failures? The most widely recognised independent view is that of the German TUV organisation using DIN standard 19250. This standard is now being adapted into IEC-1508 (still in Draft form) and by the UKOOA recommendations in principle, but the basis of categorisation is still being debated in these documents.

In effect, TUV apply a categorisation to a system, the higher the categorisation then the higher the integrity. Only one solid state system has been awarded class AK7. A number of other solid state and multiprocessor systems have class AK6 and many more have class AK5.

The categories are shown on the enclosed graphics later.

In short, when integrity is the issue, a solid state system will provide a higher level of independently assessed integrity, and the calculations of the system reliability are auditable by commonly recognised methods understandable by most people with only a basic mathematical background.

## 3.2 High Availability

There are many misconceptions about how availability can be achieved in safety systems. The most common misconception is that you can only have high availability if you have two systems voted, such that both must agree before the system shuts down. The drawback of this is that when one system fails, it may be due to an unrevealed fault as described earlier. In this case that system will remain on, and due to the voting it will inhibit the other system from tripping. In short, high availability and high safety are counter productive arguments in dual systems. The safety level of the single system is double that of a dual system in this configuration.

The next logical progression is to consider a triplicated system voted 2oo3. This does in fact give a better combination of availability and safety, but the diagnostics which define which part (or parts) are telling the truth are also a common mode failure item.

There are other combinations of the above, but the real issue is 'Why would the system fail in the first place?'. Redundancy is the result of presuming that there will be failures, that they may be hard to detect and that the periodicity of those failures will be unacceptable.

If suitable testing has been performed prior to installation, then failures occur due to an excessive change in the operating environment of the system. If the system is susceptible to small changes in its operating environment, then failures would be more likely. If the system can tolerate wide changes in its operating environment, then it is less likely to fail due to a change in temperature, or vibration, etc.

Bear in mind that operating environment is a common mode failure item for redundant systems.

If the circuit design can accept wide drifts in component parameters, that components are de-rated and that the circuit boards are built to a rugged standard, then a very high availability can be achieved with a simplex system.

The most likely points of failure are in fact the input transmitters and the ESD valves as they are more exposed to the environment. It is the input transmitter loops which are better served by triplicated redundancy. The rest of the system can be simplex. ESD valves are described later.

If it is considered that the environment subsea is as stable as you can get in terms of temperature and humidity, then the most likely cause of failure is vibration. Solid state systems by their discrete component nature are well suited to withstanding these issues.

Clearly, there are requirements for communications which rely on integrated electronics, but a failure of these communications should not have to result in a shutdown if the safety part is of sufficient integrity to be trusted to carry on with its prime function, as described later. The communications can be made completely redundant without impact on safety.

In summary, solid state safety systems meet the requirements for availability by their simplicity and tolerance to external factors. This performance has been shown on large topsides systems to be better than redundant PLC systems.

## 3.3 Redundant Inputs

As already mentioned, the most likely candidates for redundancy are input transmitters due to their exposure to the process (indirectly) and their inherent reliability.

Triplicated and voted transmitters for a single loop trip are a means to minimise the frequency of maintenance access to the transmitters. Each transmitter has a dedicated input trip amplifier which provides the trip function, and a number of diagnostics if required. The outputs of each trip amplifier are voted 2oo3 in the safety logic, with no interposing relays.

The diagnostics monitor for:

- Loop Over Current
- Loop Under Current

- Transmitter Stuck
- Process Noise

These alarms are relayed independent of the trip function to allow remote monitoring of the transmitter on alarm. Should one transmitter loop fail to a trip state then it will not generate a trip unless a second transmitter confirms it. If this does not happen then an alarm is raised. The analogue input value is also relayed via the communications link for trending topsides to detect excessive discrepancies.

When an analogue loop generates an alarm, it is important to ensure that the voting falls back to a 1-out-of-2 basis. One means of achieving this is to use a SMART Transmitter which can be programmed to fail Upscale or Downscale, thereby driving through the trip point for the input. Unfortunately, this is not guaranteed and incorrect failure modes have in fact occurred (this is a classic example of the diagnostics coverage factor for programmable systems).

In the solid state system the guaranteed fail condition is achieved by using a dedicated inherently fail safe output driver of the safety system as the power supply for each transmitter. The transmitter can then be switched off via the safety system through the communications.

Testing of the trip functions per channel is achievable by one of two methods. The first involves monitoring the response of the transmitter to methanol injection to the impulse line. This allows each transmitter in turn to be tested without tripping the logic. The second involves remotely lowering the trip level to force the input to trip. The latter is a lower integrity solution as it implies that a trip input tripping at, say, 10%, will correspondingly trip at 60%.

It may be considered that it is not only critical to know that an input will trip on demand, but that it will do so at the correct level. This may not be the case if there is a high resistance short on an analogue input which pulls the input signal down. The latter method is used on software based HIPS systems where the pressure transmitter is mounted flush with the process line.

If a fault is detected when testing an analogue input, then the ability to use output drivers as high integrity power supplies to the transmitters is an advantage of solid state systems in substantiating the system integrity where components cannot be guaranteed to fail safe.

## 3.4 Diverse From Control Functions

It is very much favoured by certifying authorities that the safety functions should be diverse from the control functions. If the two are integrated to the level that they function as a single system, then the control system could create an unsafe condition and then fail to correct for it due to common mode operations. Using diverse technology further reduces possibilities of common mode faults.

Of course, the control package may be considered to be of suitable integrity to perform the safety functions, but how do you prove this to a third party? It may be economically unfeasible to subject the control system for independent assessment as a safety system. In fact, it would probably be easier to perform the control functions in the solid state safety system.

## 3.5 Remote Communications

Operationally speaking, it makes sense to be able to find out the status of the subsea systems without the need to visit them. Should the system trip it would be nice to know why. It would be essential to be able to reset the system when the fault is corrected.

Combine these requirements for two way communications, and many people would immediately assume that solid state technology cannot support this without more bolt-on equipment.

In fact, the highest integrity solid state system has this communication capability built in and includes full isolation of the communication circuits from the safety loops. This communication can be fully redundant down to individual logic gates in the safety logic, with data processing subsea or topsides for event logging.

The communications can share an existing modem or remain dedicated through the umbilical.

Solid state systems can therefore be justified subsea in terms of the communications being redundant if desired, and more importantly the communications are isolated from the safety functions.

## 3.6 Autonomous Safety Functions

Losing communications from topsides to subsea is a pitfall of this method of oil extraction. However, there is no reason why production should be compromised unnecessarily if there is confidence that the safety functions are maintained. The use of solid state technology of Inherently Fail Safe design, with full isolation of safety functions from communications provides the confidence that even if you can't monitor the wellhead, you know that the shutdown system is continuing to monitor the sensors and will shutdown only if required.

The system can even be set to automatically shutdown if communications are absent for longer than a set time, with the watchdog function performed by the safety logic, not by a separate facility. This ability to use high integrity logic as a watchdog is a further advantage for the solid state solution.

## 3.7 Remote Testability

It is widely recognised that the greatest contributor to the unreliability of a subsea system is the ESD valve. Work is ongoing to improve valve designs to be more tolerant of the hostile process, but at present the reliability of many designs is substantiated by the test regime and periodicity. It is therefore highly desirable to be able to perform two types of test on the ESD valve:

- Partial Closure Test
- Full Closure Test

A Partial Closure Test is performed to provide confidence that the valve will move off its seat on demand, whereas a Full Closure test is to ensure that the valve closes completely.

The advantage of a Partial Closure Test is that it can be performed on-line. A request is made from the topsides to initiate the test and the valve is commanded to close. When a closure limit is met, the valve is reopened. The test status is reported via communications to the topsides.

The areas of concern during this test are the certainty that the test will complete and return the system to ready, and that the test does not override genuine trips. The validity of test requests should be verified by dual requests voted at the safety logic level.

The guarantee that a test completes can be verified by two methods, either on their own or preferably both combined. The test request is a pulse which is processed as a 'one shot' function and this is latched, this results in the output to the ESD valve solenoid(s) de-energising, thereby causing the valve to close.

The 'One Shot' function is specific to some solid state logic systems and ensures that a test request is not held on if the incoming request pulse fails to a steady 'on' position. The position of the valve can be monitored by a VPI to an analogue trip amplifier. When the valve reaches a set position the trip amplifier trips and the test latch is reset. If the VPI feedback does not occur in a set time then the test is cancelled by an inherently fail safe timer.

As can be seen on the accompanying schematic, the test request is applied in the shutdown logic after the latch for the genuine trip, such that any genuine trip will still de-energise the outputs regardless of the current test status.

The event status for all digital changes of state are logged and time-stamped to 1 millisecond resolution and analogue values are retransmitted as 8 bit words on the communication link.

A full closure test is performed by invoking a trip condition on the inputs, or by a remote input to the logic via communications. A trip on the inputs can be invoked by applying methanol injection to two impulse lines simultaneously.

## 3.8 Space

All the above issues have ignored the major constraint of subsea systems. Any system must fit in the space envelope of the electronics capsule.

Redundancy means space. Diagnostics mean space. Test circuits mean space. Diversity means space. To address this, a major advantage of solid state systems is the flexible packaging. Boards are small and modular, and can be repositioned in any orientation.

Redundancy can be limited to only those circuits which require it, diagnostics and testing are inherent features and not additional hardware.

A rack layout drawing is attached as an example.

## 3.9 Weight

Any hardware used subsea that is to be retrievable must ideally be able to be carried by an ROV. The electronics capsule incorporating control and safety systems cannot be too heavy to be brought to the surface by the ROV, so light weight components are essential. The solid state modules in aluminium racks meet these requirements.

## 3.10 Power

All power to the subsea systems must be transmitted via the umbilical. The current required by a solid state system @24Vdc to include the logic, 6 pressure transmitter supplies, 4 output solenoids and communications interface is approximately 4 Amps.

### 3.11 Cost

The small size of the solid state system ensures that cost is low without compromise to safety integrity. Designs vary little between systems as the space constraints tend to limit any overindulgence in the specifications. The systems can therefore be treated more as products, in other words a black-box. The cost is negligible in relation to design flowlines for maximum overpressure conditions.

## 4 Conclusion

Solid state technology for subsea HIPS applications provides high integrity combined with high availability, and the design and build is easily auditable by certifying authorities.

Misconceptions regarding communications limitations are unfounded.

The low cost of installing and maintaining such a system compared with the overall asset value and possible deferred production costs for nuisance trips should result in serious consideration being given to this technology as an investment, not an overhead.

## High Integrity Solid State Shutdown Systems In Subsea Locations

Eur. Ing Ian K. Ramsay-Connell
UK Support Manager
GTI Industrial Automation Ltd.

gti industrial automation

## Why?

- Pressure To Reduce Production Costs.
- Use Of FPSO Methods Combined With Subsea Completions.
- Pressure To Improve Operational Safety.
- Pressure To Provide Auditable Safety Functions.
- To Reduce Pipeline Costs By Preventing Over Pressurisation

## Topsides Shutdown

- Dependence On Comms
- Power Distribution
- Umbilical Costs

## Combined Control/Shutdown

- Less Space
- Less Hardware
- Lower Power

- No Diversity
- Safety Level?
- Common Mode Faults.
- Safety Case.

gti industrial automation

## Design Constraints

- High Integrity
- High Availability
- Redundant Inputs
- Diverse From Control Functions
- Remote Communications

gti industrial automation

## More Design Constraints

- Autonomous Safety Functions
- Remote Testability
- Space
- Weight
- Power
- Cost

## HIPS System Scope

- 6 Analogue Inputs (2 x 2oo3 Voted)
- Analogues To Include Tx Diagnostics
- 2 ESD Valves via 4 Solenoids
- 2 Analogue Inputs From ESD Valve VPI
- 2 Confirmatory Test Request Inputs Via Comms
- Handoff Off All I/O Status Via Comms
- Handoff Of Analogue MV's Via Comms

## Complexity?

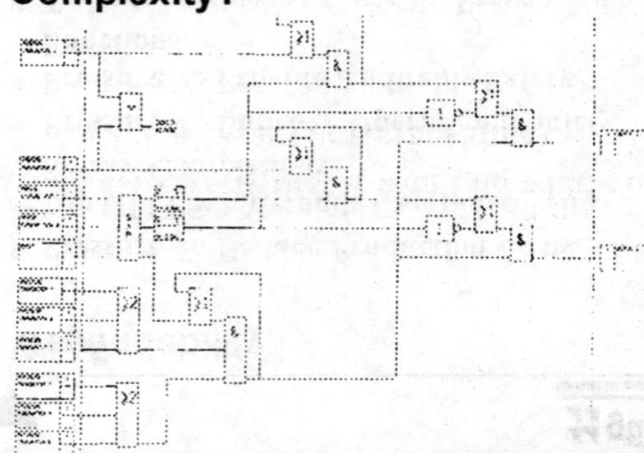

gti industrial automation

## High Integrity?

**The confidence that a system will do what you want, when you want, every time, under a wide range of conditions...**

gti industrial automation

## Dependability

**(and conversely that it won't do what you don't want any other time)**

gti industrial automation

## MIL.HDBK.217

- **Database Of Component Reliability**
- **Provides Parts Count Analysis (PCA)**
- **Provides Parts Stress Analysis (PSA)**
- **Does Not Take Account Of Circuit Design**
- **Is Pessimistic**
- **Does Not Differentiate Between Nuisance And Unsafe Failures**
- **Is A Quantitative Method**
- **Only applies To Hardware.**

gti industrial automation

## Unsafe Failures

**Those failures which are unrevealed by the system diagnostics in normal operation...**

**... and result in a failure to operate on demand when a hazard is detected**

## Coverage Factor

**The Percentage of System Failures Detectable by the Diagnostics**

**(This is probably unknown for a control system)**

## Solid State Diagnostics

- **Not an External Function**
- **Inherent In Operation**
- **Not Software Based**
- **Virtually 100%**
- **Permanently On-Line**

## Unsafe Failure Probability?

- **TUV Classification**
- **DIN 19250**
- **IEC 1508 SIL Levels**
- **UKOOA**
- **ISA SP84**

## Acceptable Risk Levels

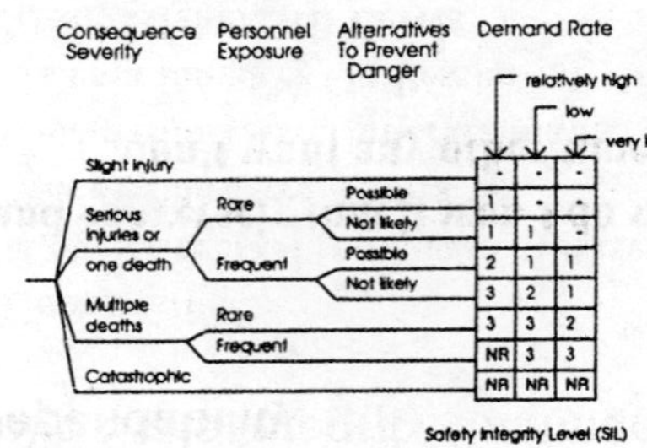


- = No special safety features required.

NR = Not Recommended. Consider alternatives.

## IEC 1508 Draft SIL Levels

| Safety Integrity Level (SIL) | Probability of Failure on Demand (PFD) |
|---|---|
| 4 | $10^{-4} - 10^{-5}$ |
| 3 | $10^{-3} - 10^{-4}$ |
| 2 | $10^{-2} - 10^{-3}$ |
| 1 | $10^{-1} - 10^{-2}$ |

## Solid State Integrity Summary

- Quantitative
- Higher Level Of Independent Assessment
- Calculations Are Auditable
- Calculations Are Understandable
- Calculations Err On Pessimism
- Meets Higher SIL Levels

## Availability

## How Is Availability Achieved?

- Fault Tolerance (Redundancy)
- Stress Tolerance (De-Rating)

gti industrial automation

## Redundancy

- Requires Voting
- Presumes Failures
- Components Fail
- System Stays On Line (for single failures)
- Requires Diagnostics
- Diagnostics Create Common Mode

gti industrial automation

## De-Rating

- Simplex (fewer parts to fail)
- Not Dependent On Component Parameters
- Tolerates Wide Parameter Drift (Low PPFS)
- Tolerates Multiple Component Drifts
- Requires Severe External Stress To Fail
- Inherent Diagnostics
- Designed For Isolation Between Circuits
- Failures Confined To Single Loops

gti industrial automation

## Why Do Systems Fail?

**There Must Be A Severe Change In The System Environment Leading To Some Part Of The System Exceeding Its Performance Tolerance.**

gti industrial automation

## Availability Summary

**Solid State Systems Meet The Demands For Subsea Applications By Their Ability To Tolerate Severe (changes in) Environmental Conditions.**

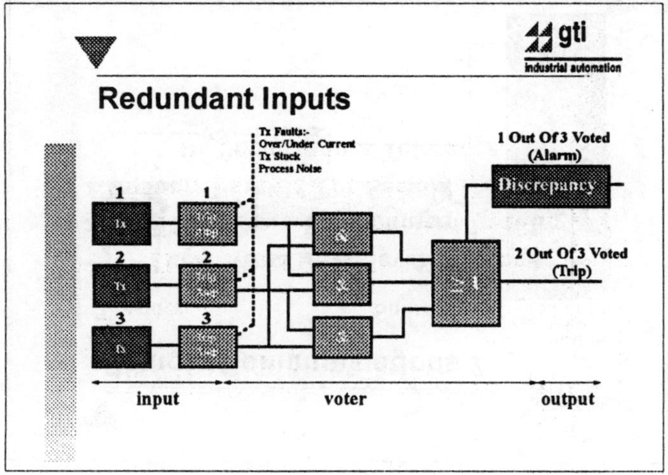
gti
industrial automation
Redundant Inputs
Tx Faults:-
Over/Under Current
Tx Stuck
Process Noise
1 Out Of 3 Voted
(Alarm)
Discrepancy
2 Out Of 3 Voted
(Trip)
1
2
3
Tx
input
voter
output

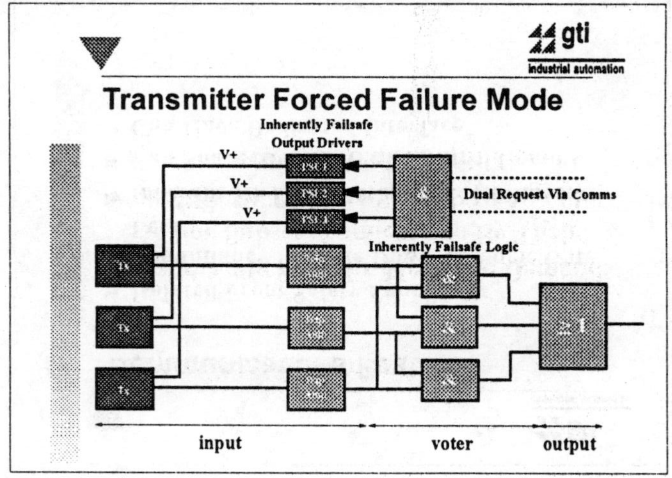
gti
industrial automation
Transmitter Forced Failure Mode
Inherently Failsafe
Output Drivers
V+
V+
V+
Dual Request Via Comms
Inherently Failsafe Logic
Tx
input
voter
output

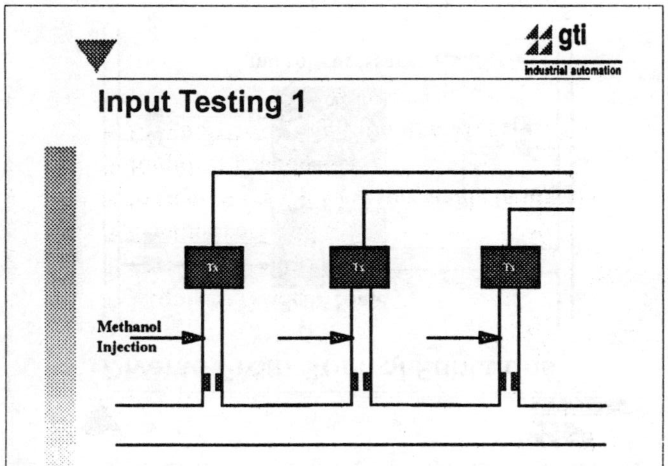
gti
industrial automation
Input Testing 1
Tx
Methanol
Injection

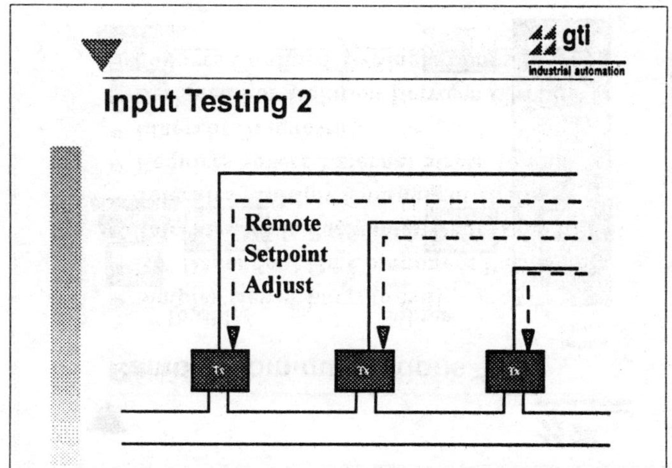
gti
industrial automation
Input Testing 2
Remote
Setpoint
Adjust
Tx

gti industrial automation

## Diverse From Control Functions

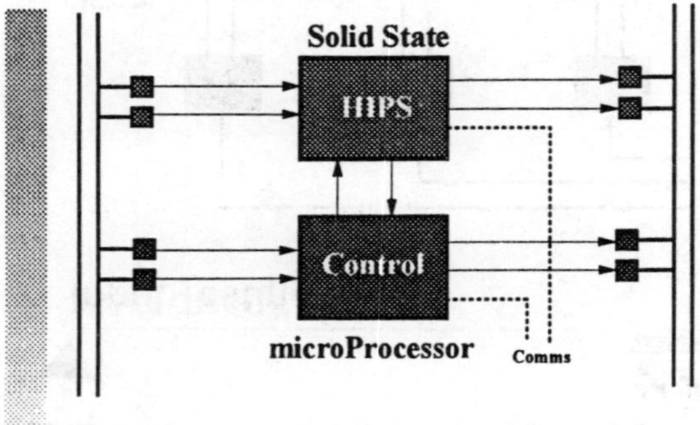


gti industrial automation

## Remote Communications 1

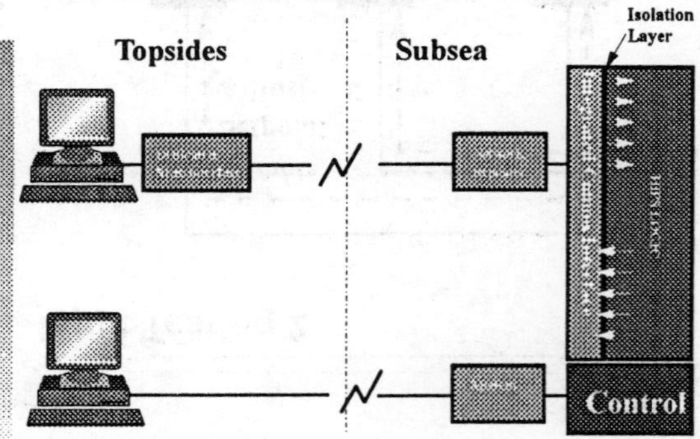


## Remote Communications 2

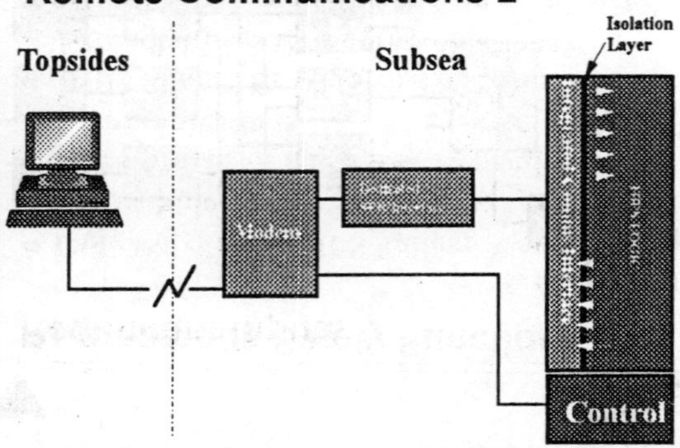


gti industrial automation

## Communications Features

- Isolated From Safety Functions
- Redundancy Possible Down To Logic Gate Level
- Inherent High Resolution SER (to 1msec)
- Can Share Modem With Control Comms
- Can Have Dedicated Interface

## Autonomous Safety Functions

- Only Possible If HIPS Failure Modes Are Predictable
- HIPS Must Not Have Any Dependence On Communications
- HIPS Should Be Able To Time Out On Extended Loss Of Communications
- HIPS Should Provide The Comms Watchdog
- Watchdog Timer Should Be Fail Safe

## Remote Testability

- Instigate Test Via Comms
- Partial Closure Test (on line)
- Full Closure Test (off line)
- Test Does Not Override Actual Trip Demands
- Auto Removal Of Test Via Feedback and/or Timer
- Recording & Handoff Of Test Results

## Test Circuit

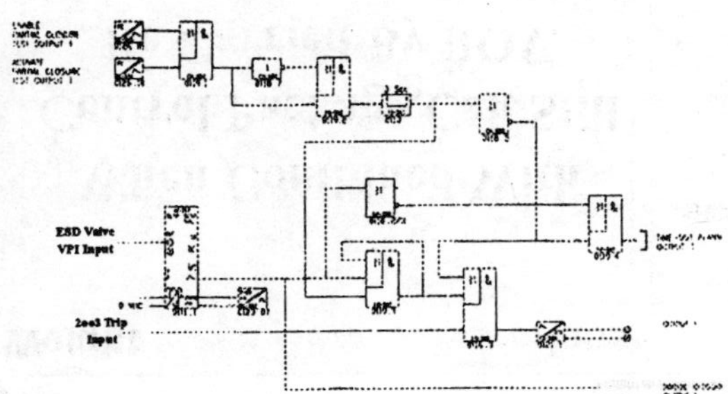


## Space

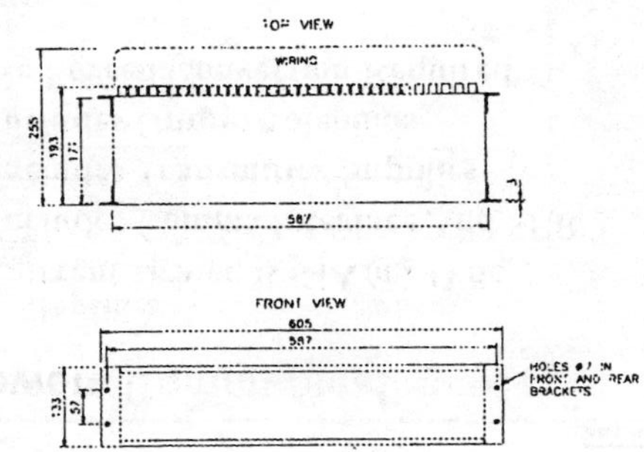

gti industrial automation

## Weight

**When Combined With Control Package,Can Still Be Carried By ROV.**

gti industrial automation

## Power

- Current Drawn is <4A @24Vdc
- Includes Comms Interface And SER
- Includes Transmitter Supplies
- Includes Output Solenoids
- No Forced Convection Required

gti industrial automation

## Cost

**Considerably Less Than The Cost Of Having A Flow Line Built To Withstand Over Pressures...**

**& Less Than The Cost Of Having A Control System Assessed And Audited For Safety Applications.**

gti industrial automation

## Conclusion

**If A HIPS Is required For A Subsea Package, Solid State Systems Can Provide The Highest Required Integrity & Availability Without Compromising On Functionality And Communications,**

**And Without Excessive Cost Implications**

gti
industrial automation

## Only The System Should Be In Deep Water!

# The development of subsea high integrity pipeline protection systems (HIPPS)

L A ADRIAANSEN
DNV

Synopsis:

This paper describes the design and construction of a subsea "High Integrity Pipeline Pressure Protection System" (HIPPS), which enables subsea pipelines to be designed for the operating, rather than the shut in wellhead pressure. Such systems will save considerable investment cost in the development of high pressure offshore oil and gas fields, particularly where the distance to the processing infrastructure is long.
This safety system comprises two rapidly closing gate valves in a double block and bleed configuration to protect the pipeline from over-pressure. The reliability of the system is assessed and dynamic simulation of the valve and process flow are discussed.
The paper demonstrates that the operational safety criteria can be met and that the proposed design is technically feasible from both a safety and operational point of view.

L.A.Adriaansen BA Eng., DNV

L.A.Adriaansen studied Marine Engineering in the Netherlands and Control and Instrumentation in the UK. After a career start in Marine Engineering he joined Selection Trust Ltd and became engaged in the mechanical engineering design of extractive metallurgical processing plant and later the design of the process control systems for such installations. After joining BP, he became responsible for subsea production control systems. Presently employed by DNV, where he was head of the Subsea Production System Department, he is now seconded to Kongsberg Offshore a.s. and responsible for the development of a subsea HIPPS installation for Statoil.

## INTRODUCTION

### General

A number of subsea field development studies, in particular for marginal fields, have identified that the flow and pipeline cost represents a significant portion of the total field development cost. The pipeline cost is of course also largely dependent on the reservoir pressure. In some cases where the reservoir pressure was high, the studies clearly showed that the field could not be economically developed due to the high pipeline cost.

It was also clear from these studies that if the pipeline could be designed for the Maximum Operating Pressure (MOP) instead of the Shut-in Wellhead Pressure (SWHP), many of the developments would be economically feasible.

This background has generally encouraged the view that a pipeline safety system would be required for the development of high pressure marginal fields. This would then permit transportation of hydrocarbons at reduced pressures. Such a safety system must then ensure that the pipeline pressure can not exceed the design pressure, under any circumstances. Such systems are now being seriously considered and have become generally known as "High Integrity Pipeline Protection Systems", HIPPS for short.

A safety or protection system should be of high integrity and reliability. Apart from land based application, such a system has already been implemented on an unmanned offshore platform in the North Sea (Sleipner West), to protect the export pipeline (32" NB, 20 km) to the Sleipner B platform.

A HIPPS system for subsea application, although principally identical in function to the land based or platform based system, should have even higher reliability and most certainly higher availability due to the difficulty and cost associated with intervention.
The probability of hydrate formation and subsequent blockage of the pipeline must be considered. In addition, the special problems associated with a subsea environment would have to be solved.
A subsea HIPPS system should also be able to cope with both gas and oil fields (high GOR) of high pressure.

<u>Design basis</u>

In early 1994, work started on identifying the problems associated with subsea HIPPS applications and to establish the scope of work and detailed technical requirements for the development of such a system for Statoil using the following design criteria as input:

- Field proven gate valves should be used.
- Field proven technology should be used.
- Dual redundancy was required.
- The "trigger" system should be based on a voting system.
- The system should be autonomous and independent of the well control system.
- In situ testing of the valves should be possible.
- Simplicity, reliability and availability should be key considerations.
- The system should be suitable for a water depth of 500 metres.
- The acceptable failure rate of the system should be less than $10^{-4}$.

Some of the arguments used in arriving at the design basis are discussed below.

## SYSTEM CONFIGURATION

<u>HIPPS system selection</u>

In designing a subsea HIPPS installation, many options are available to the designer to achieve his objectives. Some options are better than others, whereas others suit the cultural climate of the operating company better.

This section attempts to discuss some of the relevant issues, which are primarily associated with the following:

- Reliability/availability
- Number of HIPPS systems per subsea installation
- Type and number of valves
- Fail safe mode
- Actuation of the valve
- Type of control fluid
- Use of Xmas tree valves
- Use of the production control system
- Trigger mechanism
- Redundancy
- System autonomy

***Reliability/availability***

When addressing reliability and availability of a safety system, it is perhaps better to think in terms of "safety performance" and "production regularity performance", both of which are defined below.

Safety performance:

"Safety performance" is the probability that the system fails to operate upon demand, i.e. the probability that the system does not respond to a manifold pressure in excess of a pre-set level and fails to close the HIPPS valves. The probability of failure to operate upon demand is denoted as "Critical Safety Unavailability" (CSU).

Production regularity performance:

"Production regularity performance" is meant to mean an assessment of the number of spurious activations of the system per unit time, e.g. the closing of the HIPPS valves in the absence of too high a pressure in the manifold. The production regularity performance is denoted as the "Spurious Trip Rate" (STR).

To improve the CSU a degree of redundancy has to be built into the system. This increases the number of components and hence the STR. There may thus be a potential conflict between the CSU and STR.

The availability of the system is a function of the repair time. However, if the level of redundancy in the system is sufficient, repairs should not be necessary during the service life of the system. In this respect it is important to bear in mind that the HIPPS safety system may, depending on the specific reservoir characteristics, have to be available for only a relatively short time compared with the production system.

Recent studies undertaken have indicated that the service life of a HIPPS system may typically be from a few months to 5 years. On completion of the service life, the SWHP is deemed to have dropped sufficiently below the pipeline design pressure.

***Number of HIPPS systems per subsea installation***

For a subsea production installation, whether it is a manifold/cluster or template scenario, one can only conceive the following configurations for a typical four slot installation:

1) Single HIPPS per installation:

+ Approximately fourfold increase in reliability compared with 2 below.
- Total production loss with a spurious trip.

2) Single HIPPS per well:

+ Single well production loss only with a spurious trip.
- Approximately fourfold decrease in reliability compared with 1.
- Requires the added complication of a dedicated subsea hard wired telemetry link between all 4 systems to ensure that all 4 systems trip simultaneously.

3) Single HIPPS per group of wells:

The arguments for this case are self explanatory and may be based on the information above.

***Type and number of valves***

When safeguarding pressure containing equipment against over-pressure through the use of pressure relief valves, always two independent devices are employed. It would therefore be prudent to use the same philosophy in the design of a subsea HIPPS system. Instead of relieving the pressure however, the pressure source is blocked in by means of two independent barrier valves in series.

The type of valve used is rather immaterial, providing the valve meets the technical requirements in respect of operability, functionality, leak tightness and reliability. Should "pigging" be a requirement, then the choice of valve is typically restricted to ball valves and gate valves. Gate valves have however demonstrated to have better sealing characteristics than ball valves. The latter being more bulky but occupying less height overall.

***Fail safe mode***

The failure mode of the valve must be "fail safe" on loss of electrical or hydraulic power. "Fail safe" implies that the valve fails in one of the following modes:

- Fail "stay put" (or "as is")
- Fail open

- Fail closed

Fail "stay put"
If the valves fail "stay put", then the most likely end state is the "open" position, as the HIPPS valves are open most of the time. The pipeline would then not be protected.

Failure of the hydraulic supply can be monitored and alarmed. Failure of the solenoid drive circuit and coil of the directional control valve can also be monitored and alarmed. The operator will thus know that the system is no longer operational and can no longer protect the pipeline. Operator intervention is thus required to initiate appropriate action. This is undesirable for this type of safety system, as it may lead to error of judgement by the operator, under pressure to meet production quotas.
If the system is not operationally available, then production must cease to avoid the risk of damage to the pipeline. This should be an automatic function.

Fail open
If the valves fail "open", then the same arguments used for the "stay put" case apply.

Fail closed
The only acceptable failure mode must therefore be "fail closed", as this ensures that the pipeline remains protected and that appropriate action has to be taken to restore the "health" of the safety system.

***Actuation of the valve***

The energy required to ensure that the valve "fails closed" may be derived in 4 ways; mechanical spring force, hydraulic energy, process pressure or a combination of these.

A hydraulic actuator with a hydraulic energy source will always be required for subsea applications to drive the valve either open or closed, ignoring the concept of "electrical actuators".

When using a reverse acting gate valve, the process pressure will tend to drive the valve closed, a principle known as "pressure assist to close". This means that the actuator requires hydraulic power to keep the valve open.
This is in contrast to a ball valve where one could conceivably argue that the valve will remain in its last position on loss of hydraulic power.

In gate valve applications, a mechanical spring is used to ensure that the valve will close.
In a HIPPS application, one may reasonably assume that process pressure is always available.
If the process pressure is not available or insufficient, the HIPPS valves are not required to function. This means that the process fluid pressure could be considered as an energy source to drive the valves closed.

There is no inherent requirement for supplementary force (i.e. spring, etc.). By deleting the spring and using the process fluid as an energy source, the HIPPS space "envelope" and the number of components may be significantly reduced, leading to increased reliability.

*Type of control fluid*

The types of control fluids available for this application are basically: water based fluid, mineral fluids and synthetic fluids.

It is beyond the scope of this paper to enter into a detailed discussion on the merits of these fluids. Suffice it to say that a HIPPS system should be able to function with any of these fluids and that the fluid selection should be based on that to be used for the production control system.

It should however be borne in mind that water based fluids may be discharged to sea, whereas the other fluids have to be returned to the surface (back pressure in the system can thus be expected).

*Use of Xmas tree valves*

Having argued that two valves are required to "block in" the pressure source, one could conceivably argue that the wing and master valves on the Xmas tree are available to undertake this role.

Regardless of the issue of how the valves should be "triggered", it is technically feasible to use the wing and master valve, providing that these valves can be closed rapidly. Normally, such valves close in 20 to 30 seconds. Modifications would thus be required to the hydraulic systems to achieve faster response times. It would most probably also be necessary to close down the other well systems and a complex control interconnection network might ensue.

It should not be overlooked that the primary role of the HIPPS system is to ensure safety. Such systems should normally be independent and thus not form part of any other system. The HIPPS valves should also be kept in a pristine condition to function correctly when required to do so. Master valves on conventional trees on the other hand, can sustain considerable damage through "gauging" operations and the passage of various downhole tools.

The issues involved are not clear cut. They are subjective and most probably influenced by cultural thinking of the operators and regulatory authorities.

*Use of the production control system*

If the Xmas tree valves are considered for a HIPPS application, then logical extension of the argument will focus attention on the production control system as a potential trigger system.

Again, it is technically feasible to incorporate the trigger systems in the Production Control Module (PCM) electronic systems, but not without penalty.

Subsea production control systems are primarily designed to operate and monitor the subsea installation. Although very reliable, the integrity of such systems in general, may be considered insufficient to be part of a pipeline safety system. The production control system hardware would thus most likely have to be upgraded to achieve the same degree of integrity as is required for surface based safety systems.

Safety systems, on the other hand, should be as simple as possible. The simpler a task and the fewer components used to ensure execution of the task, the higher the overall system reliability. The PCM software has to perform many different tasks that one would not wish a safety system to undertake. A safety system should be dedicated to the safety task only.

If a single HIPPS system is used in a subsea installation and if the associated trigger system is located in a single PCM, then the safety system would not be available if that PCM has been retrieved to the surface for maintenance. It thus follows that to maximise the availability of the HIPPS system, a dedicated control module is preferred.

***Trigger mechanism***

The arguments used in deciding to use a solid state or software based safety system are classical and beyond the scope of this paper.

The use of microprocessor based systems in surface safety systems is well established. Providing that the same degree of integrity is applied in the design of the subsea system, then an acceptable degree of safety and reliability may be realised.

The trigger system requires an input from pressure transmitters. The number of pressure transmitters required is debatable, but an acceptable level of redundancy must be realised.

The manner in which the trigger system treats the inputs requires careful scrutiny. A voting system seems a natural choice, but the manner in which the voting is carried out should be simple and resistant to errors. "Out of range", "trend" and "self" checking procedures should be an inherent feature of such a design. Common mode failures should be avoided at all cost through careful design methodologies.

Although a software based system has been selected, a solid state based system could be implemented without difficulty.

***Redundancy***

Having accepted that two HIPPS valves are required, it is only prudent to ensure that the trigger electronics are also duplicated and that no interaction between the two systems leads to common mode failures.

This level of redundancy should be carried through to the sub-units in the control system. Only components that, through their inherent simplicity and field proven track record are deemed to be ultra reliable, do not require duplication.

***System autonomy***

An inherent feature of almost any safety system is that the system is autonomous. This means that the system decides for itself, on the spot and without operator interference, what action should be taken. A non-autonomous system would be inappropriate for this type of safety application as it would be inconceivable to require the decisions to be made on the surface and rely on a telemetry link to convey the desired commands.

## Design rules

To be effective in the design of this safety system, it was decided at an early stage to create a set of design rules based on some of the arguments advanced above. In doing so, a mechanism was thus established to ensure that all involved in the design were fully familiar with the general requirements and could be monitored in respect of achieving the desired objectives.

The more significant of these rules are set out below:

- The HIPPS system shall be a stand alone autonomous system.
- The safety performance of the HIPPS system shall be as high as possible, CSU < $10^{-4}$
- Every effort should be made to reduce the probability of spurious trips.
- The HIPPS system shall have its own dedicated control module.
- On detection of over-pressure, the HIPPS system shall simultaneously close two valves mounted in series.
- The HIPP's telemetry link to the surface shall have its own dedicated set of conductors in the production control umbilical.
- The system shall be designed to be tamper proof, this also applies to the set point of the initiating device (trigger), which may be software based.
- It shall not be possible to "down load" critical safety software.
- Critical safety software shall be resident in EPROM.
- The pressure in the cavity between the two valves shall be monitored and be capable of being vented into the service or annulus vent line via a dedicated bleed valve.
- It shall be possible to flush the process connections of the pressure transmitters located between the valves with methanol.
- It shall be possible to "in-situ" test the pressure transmitters initiating the HIPPS valve shut down.
- The pressure transmitters shall have a process connection of at least 1" NB and be mounted as close as possible to the process stream.
- The HIPPS system design shall have the least number of common mode failure points.
- The HIPPS valve closing time, from detection of a high pressure to the valve having reached the closed position, shall be < 2 seconds.
- The valve design shall be tolerant to impact loading due to rapid closing.
- The number of components in the system shall be kept as low as possible.
- The HIPPS components in the system with identical functions may be identical equipment from the same supplier, subject to satisfactory field performance.
- Regular in-service testing shall be required in accordance with the statutory requirements laid down for the down hole safety valve.

- Each HIPPS valve shall have its own "trigger" system. Once triggered, both valves shall be closed.
- The HIPPS system shall be retrievable to the surface for repair or replacement.
- The HIPPS control module should be separately retrievable.
- Once the reservoir pressure has sufficiently decayed, a simple "lock open" feature should eliminate spurious trips.
- The required failure mode of the HIPPS valves shall be "fail closed".
- The Directional Control Valves (DCV) shall generally be of the electrically pulsed, hydraulically latching type.
- Two series mounted DCVs shall be used for each HIPPS valve.
- Electrical continuity monitoring shall be applied to the DCV coil circuits.
- The "Trigger" function is to be arranged in software with a 2 oo 5 voting system (2 out of 5), which progressively degrades to a 1 oo 1 voting system on detection of transmitter failure.
- Once the HIPPS valves have been "triggered" to close, it shall not be possible to open the valves until a dedicated software or hardware "reset interlock" (key switch) in the surface control system has been reset.
- When the "Trigger" has operated, a priority message shall be sent to the surface to take appropriate automatic action in the ESD system.
- In "test mode", it shall be possible to open and close the valves individually via a hardware or software "test mode" switch in the surface control system.
- Diagnostic and self checking features shall be incorporated in the design. Such features shall not impair the reliability of the system.
- Position transducers shall be fitted to the HIPPS valves to determine valve position and the opening and closing characteristics.
- The HIPPS safety valves shall close on loss of electrical power.
- The system shall remain functional on loss of communication.
- It shall be possible to individually open each HIPPS valve under hydraulic power.
- The HIPPS system may derive its hydraulic supply from the production control system.

In establishing some of the above design rules, detailed reliability calculations were performed to ascertain the optimum system configuration from the options available. This was particularly relevant for the voting system, pressure transmitter configuration, "trigger" electronic configuration and the hydraulic system configuration.

## SYSTEM DESCRIPTION

### General

The HIPPS system (see fig 1) comprises two rapidly closing (2 sec), series mounted valves, an electronic "trigger" system and five pressure transducers. The latter provide the input to the trigger system.

The "trigger" system is autonomous and is located in a dedicated (retrievable) control module, which uses the same hydraulic power system as the subsea production control system. Electrical power is arranged via a dedicated power supply in the production control umbilical.

The system has been designed to provide ultra high reliability and availability through careful system evaluation and component selection. In this process, every precaution has been taken to avoid the possibility of spurious trips.

## Trigger system

The trigger system (see fig. 2) comprises two independent micro processor based systems which receive as input the signals of 5 pressure transmitters. Three pressure transmitters are located between the HIPPS valves and two are located down stream. The output of the trigger is connected to the solenoids of the directional control valves in the hydraulic circuit. The pressure transmitters are equipped with dual electronic measurement circuits to achieve total separation between the two systems. The trigger uses a 2-out of-5 voting system which progressively degrades to a 2 oo 4, 2 oo 3, to a 1 oo 2 voting system on detection of transmitter failure. This process is schematically represented in fig 5. Each trigger receives electrical power via dedicated conductors in the production control umbilical. The telemetry signals are superimposed on the electrical power circuits. The triggers are therefore capable of communicating with the surface facility and can receive commands such as required during periodic testing of the system. Internal system checks and electrical continuity checking ensure that the system is in as high a state of preparedness as possible. Each trigger circuit is located in a one atmosphere nitrogen charged container located in the dielectric fluid filled HIPPS control module, which also houses the hydraulic components.

## Hydraulic system

The hydraulic system is schematically shown in fig. 3. The system has the lowest possible number of components commensurate with the objectives of simplicity and reliability. To achieve the latter, dual Directional Control Valves (DCV) have been incorporated in series to optimise the reliability of those DCVs forming part of the safety function.
The DCVs are of the well proven shear-seal type, electrically pulsed and hydraulically latching. The DCVs have dual solenoid coils, separately connected to each solenoid drive circuit in each trigger to ensure complete circuit separation.
The two series mounted DCVs are dedicated to the safety function and are connected to two 1" gate valves, one mounted on the actuator of each HIPPS safety valve. The "fail safe" coils of one of these valves may be continuously energised to ensure HIPPS safety valve closure on loss of electrical power to both circuits, should this be required.
The purpose of the 1" (fail open) gate valves is to create a hydraulic interconnection between the volumes above and below the pistons in each actuator, hence the name "Xover" valve. Once the Xover valves are opened, the control fluid in the actuator is rapidly transferred from the "opening" side of the piston to the "closing" side, enabling HIPPS safety valve closure in less than 2 seconds.

Pressure transmitters in the circuit are for "house keeping" purposes. Valve "signatures" are obtained of the safety valves each time the valve is stroked by means of the positions sensors located in the rising stem.

A water based control fluid has been selected for this application, but the system will function equally well with a synthetic or mineral control fluid. An accumulator in the circuit ensures that the system pressure can not drop too low during opening of the valve.

## HIPPS valves

HIPPS valves should be "fail closed" and can be any type of valve. Conventional, field proven, reverse acting, bi-directional gate valves have been selected for this application, as gate valves are considered to have superior sealing capability. In addition, bi-directional valves mounted directly on pipeline headers would remove the need for separate manifold isolation valves for some applications.

The size of conventional gate valves may however be prohibitive for some subsea applications. This prompted an evaluation of how a conventional gate valve could be re-designed to provide a more compact and cost effective solution. In this evaluation process, the design was optimised to reduce the number of components and thus improve reliability. All these objectives were met and the outline design of the valve is shown in fig. 4.

Fig. 4 shows the HIPPS optimised valves arranged in a monoblock configuration, with the bleed valve and methanol injection valve mounted on a separate block to the side of the main block. This arrangement enables the process bore cavity between the valves to be vented or pressurised to test the pressure transmitters and valve sealing performance.
Three pressure transmitters are located on the main block between the safety valves. The measurement diaphragms are flush mounted as close to the main process bore as feasible whilst reducing the risk of sand erosion. The measurement area may be flushed with methanol.

Fig. 4 shows that a conventional field proven gate valve design has been applied. The "pressure assist to close" feature of the reverse acting gate valve has been utilised to delete the spring. The process fluid thus provides the energy to drive the valve rapidly closed. Towards the end of the valve closing stroke and in particular near the "pinch point", the seal friction builds up and the amount of energy transferred to the body cavity is reduced, thus naturally arresting the valve gate motion. The actuator is of the "rising stem" type and incorporates a position transducer of the LVDT (Linearly Variable Differential Transformer) type.

Clearly, the rising stem creates a water depth limitation. The design as shown is suitable up to a water depth of 1000 metres. The water depth capability can however be easily increased by incorporating a balancing stem on the opposite side of the gate body or by adding an external spring element.

Once the reservoir pressure has sufficiently decayed, the services of the HIPPS system are no longer required. It would then be advantageous to lock the valves permanently open and thus totally avoid the possibility of a spurious trip. This may be achieved by a ROV or diver, inserting a simple key to trap the rising stem and thus the valve, in the open position.

## OPERATOR INTERFACE

The operator interface comprises typically a node in the main process control system. There are 3 operational modes available to the operator to choose from:

- The Auto mode
- The Manual mode
- The Test mode

### AUTO MODE

In this Autonomous/Automatic mode, the HIPPS system functions as a safety system with the subsea trigger set at a fixed value of say 170 bar. Should this pressure be exceeded, the HIPPS safety valves close rapidly (< 2 secs.) and the topside ESD system is activated. This mode can only be entered when the HIPPS safety valves have been opened in manual mode (reset). The methanol injection valve is operational in this mode.

### MANUAL MODE

In this mode, the HIPPS safety valves may be individually opened and closed. The trigger is functionally disabled but will cause an alarm should the process pressure exceed the set point of 170 bar. The bleed valve and methanol injection valve are operational.
This mode should only be entered when production has stopped.

### TEST MODE

In this mode, the HIPPS system is functionally tested by reducing the trigger set point progressively until the trigger is activated and the HIPPS safety valves are closed. Once the HIPPS safety valves are closed, the system has to be switched to the manual mode to open the valves in sequence in a controlled manner. This procedure may be carried out without stopping production.
The surface ESD system will not be activated.

## SYSTEM DISCUSSION

### Pipeline transient behaviour

The main purpose of the HIPPS system is to act as a safety system to protect the pipeline from over-pressure. Over-pressure may be caused by the following incidents:

- o Operator error
- o Choke valve failure
- o Unintentional downstream valve closure
- o Hydrate plug formation

Of the above incidents, hydrate plug formation relatively close to the wells is considered to be the most serious situation, as it requires the fastest response. If such an event can be safely handled by a HIPPS system, then all other perceivable events can be safely handled. Analysis of pipeline transient behaviour in respect of rapid hydrate plug formation has indicated that the pressure build-up is relatively fast, even in comparison with the rapid closing of the HIPPS valves in 2 seconds.
Line blockage due to hydrate formation is therefore considered to be the most serious event on which further analysis will be based.

One has a certain degree of freedom in deciding the margins between the pipeline design pressure, the operating pressure and the trigger pressure. Increasing the design pressure will increase the pipeline cost. If the trigger pressure is too close to the operating pressure, the number of spurious trips as a result of transient behaviour, may increase. It is therefore important to gain a good understanding of transient behaviour during normal operation. The following values may be considered as typical but would have to be calculated for each scenario:

| | |
|---|---|
| Normal operating pressure | 160 bar |
| HIPPS trigger pressure | 170 bar |
| Max. Operating pressure (MOP) | 180 bar |
| Design Pressure (DP) | 180 bar |

If a hydrate plug suddenly formed in the pipeline, then the pressure will continue to rise until the HIPPS valves have closed. The immediate section downstream of the HIPPS valves can therefore be subjected to a pressure in excess of the pipeline design pressure. This section of pipeline will therefore have to be designed for the shut-in wellhead pressure (SWHP) and is referred to as the "fortified" zone. The length of the fortified zone is field specific and has to be calculated for each case.

If it is assumed that the rate of flow through the HIPPS safety valves remains constant and that the fluid properties do not change, then the following formula may be used to get a good approximation of the length of the fortified zone.

$$L = \gamma_o \times v \times dt \times c^2 \times (P_t - P_o)^{-1}$$ where

$L$ = length of fortified zone, m
$\gamma_o$ = fluid density, $Kg.m^{-3}$
$v$ = velocity, $m.s^{-1}$
$dt$ = valve closing time, seconds
$c$ = velocity of sound in fluid, $m.s^{-1}$
$P_t$ = fluid pressure at time t, $N.m^{-2}$
$P_o$ = fluid pressure at time t = o, $N.m^{-2}$

It may thus be seen that for a given installation, the length of the fortified zone is primarily a function of the margin between the pipeline design pressure and trigger setpoint as well as the valve closing time.

This is illustrated in fig.8, which shows the relationship between the hydrate plug location and the pressure setpoint of the trigger.The final pressure reached as a function of hydrate plug location is shown in fig. 9.

## Spurious HIPPS valve closure

Spurious closure of the HIPPS valve must be kept to as low a level as possible. This requirement is to some extent in conflict with the requirements of high reliability and is further discussed under "Reliability".

In order to distinguish between real and spurious closure, adequate process monitoring, trend recording and "house keeping" routines have to be included to provide operating staff with essential information. These routines have been incorporated in the design. In addition, detailed operational procedures will be prepared to assist the operator in fault analysis and testing of the system.

## Effects of potential valve leakage

Valve leakage can be a serious problem if the rate of leakage is significant or if the leakage persists for a long period. In assessing the consequences of valve leakage, two scenarios have to be considered; leakage detected during periodic system testing and leakage detected on closure of the HIPPS valves as a result of high pipeline pressure.

Once the trigger has commanded the HIPPS valves to close, a priority message will be sent along the telemetry link to the surface control system to take appropriate action in the ESD system. The consequence of this is that the production control system will be closed within a relatively short time, typically 30 seconds. This ensures that the possibility of over-pressurising the pipeline is minimised. In such an event, the operator will follow detailed operational procedures and bleed off the cavity between the HIPPS valves, into the annulus vent line. Once the bleed valve is subsequently closed again, any pressure build up will be an indication of HIPPS valve leakage. The risk of hydrate formation during this process is small, due to the relatively small volumes involved and the presence of methanol.

If leakage is detected during routine periodic system testing, the rate of leakage will have to be assessed. To this effect, detailed operational procedures must be prepared which will enable operating staff to determine the extent of leakage by venting the process bore between the valves and observing the pressure behaviour. In addition, the cavity can be pressurised with methanol.

Periodic testing is recommended to occur at the same frequency as that for the downhole safety valve. This means for the Norwegian sector that the system is tested once a month until sufficient confidence has been gained, then every three months (until sufficient confidence has been gained) and from then on every six months.

It should be realised however that gate valves when tested in the factory are "bubble tight". In operation one would expect some leakage eventually to occur, but this should be seen as "drips" rather then significant flow rates. Recent testing of an identical 10K valve demonstrated during API 6A blow-down tests that after 846 cycles with full differential pressure no measurable deterioration in valve seal performance had occurred but that some minor valve stem packing leakage occurred during the elevated temperature tests (400 deg. C).

## DYNAMIC SIMULATION OF HIPPS VALVES

The relatively unique features of the HIPPS valve design required a thorough analysis in respect of the behaviour of the valve.
A dynamic simulation was therefore carried out using the Cypros ESIM programme. The analysis demonstrated that the closing times required are feasible with the present unmodified design up to a water depth of 1000 meters and that the dynamic forces involved are of a very manageable proportion. The water depth capability can be easily increased as discussed previously.

Fig. 6 shows the valve closing time characteristic for a water depth of 500 metre and a process pressure of 170 bar. After a slight delay during which the pressure in the actuator decays, the velocity increases rapidly from the open position, stabilises and reduces rapidly beyond the valve "pinch" point at 1.2 secs. At the "pinch" point the valve is closed but has not completely travelled yet to its end position. The latter is achieved at a continuously reducing velocity.

Fig. 7 shows the effect of decreased process pressure on the valve closing speed.

Fig.8 shows the effect on closing speed when the water depth is increased from 500 metre to 1200 metre. As the water depth increases, the pressure increases on the rising stem with the result that the valve speed reduces. At 1200 metre water depth, the valve does not completely close. The limiting water depth for the design shown is thus 1000 metres but it should be born in mind that the data shown is pessimistic due to the friction coefficients used..

## RELIABILITY ANALYSIS

In order to assess the reliability of the system in respect of the Critical Safety Unavailability (CSU) and Spurious Trip Rate (STR) defined earlier, it was decided to employ the PDS (Reliability and Availability of Computer-Based Process Safety Systems) method.
This methodology, developed in Norway, takes into consideration the potential conflict between safety and production regularity and provides the means to balance a system configuration in respect of these two performance criteria.
This is achieved by allowing the designer to evaluate the effect of various fault tolerant (e.g. redundancy and self test) and fault removal (e.g. preventative and corrective maintenance) techniques. The effects of normal ageing are considered as well as those due to environmental stresses and human factors, during design and operation.

The reliability analysis was carried out by independent authorities and yielded the following results for the total HIPPS system:

| | |
|---|---|
| Critical safety unavailability (CSU) | 0.0089 % (test interval 3 months) |
| Failure rate | $0.89 \times 10^{-4}$ |
| System availability | 99.99 % |
| Spurious trip rate (STR) | 4.36 per $10^6$ hours |
| Years between spurious trips | 26 years |

The CSU value is of course dependent on the test interval period. Each time the system is tested, the system may be considered to have started "afresh" again, even if potential critical failures have not been discovered. The test interval period is not taken into consideration in calculating the STR. The latter is judged to be insignificantly effected by the test frequencies under consideration.

If the test interval is increased to every 6 months, then the following applies:

| | |
|---|---|
| Critical safety unavailability (CSU) | 0.0159 % (test interval 6 months) |
| Failure rate | $1.59 \times 10^{-4}$ |
| System availability | 99.99 % |

## COST/BENEFIT ANALYSIS

A cost benefit analysis was undertaken to determine the likely cost savings in a theoretical field scenario.
The savings for both a 680 bar and 450 bar SWHP, Duplex system compared with a 200 bar system, may be summarised as follows:

| Pipe ID in | CIWP bar | Cap.cost NOK/m | Install.cost NOK/m | Cost NOK/m | Saving MNOK/km |
|---|---|---|---|---|---|
| 7" | 200 | 1992 | 5600 | 7592 | ------- |
| | 450 | 4685 | 11450 | 16135 | 8.54 |
| | 680 | 7162 | 16800 | 23962 | 16.37 |
| 8" | 200 | 2629 | 5600 | 8229 | ------- |
| | 450 | 6282 | 11450 | 17732 | 9.50 |
| | 680 | 9643 | 16800 | 26443 | 18.21 |
| 10" | 200 | 3842 | 5600 | 9442 | ------- |
| | 450 | 8568 | 11450 | 20018 | 10.58 |
| | 680 | 12915 | 16800 | 29715 | 20.3 |
| 12" | 200 | 6115 | 5600 | 11715 | ------- |
| | 450 | 13027 | 11450 | 24477 | 12.76 |
| | 680 | 19386 | 16800 | 36186 | 24.47 |

The analysis was based on a CIWP of 680 bar. The 450 bar case was derived afterwards by interpolation. Above figures are based on a theoretical field scenario and must therefore only be considered as indicative of the significant savings that can be made.

## CONCLUSION

The work performed sofar, has demonstrated that this pipeline protection system meets the technical and operational requirements in respect of safety, reliability, response and waterdepth capability specified.

The development of this system, at the time of writing this paper, had not entered its test qualification program yet, which is scheduled to be completed at the time of this conference. Any lessons learned during the test qualification program will be declared during the conference.

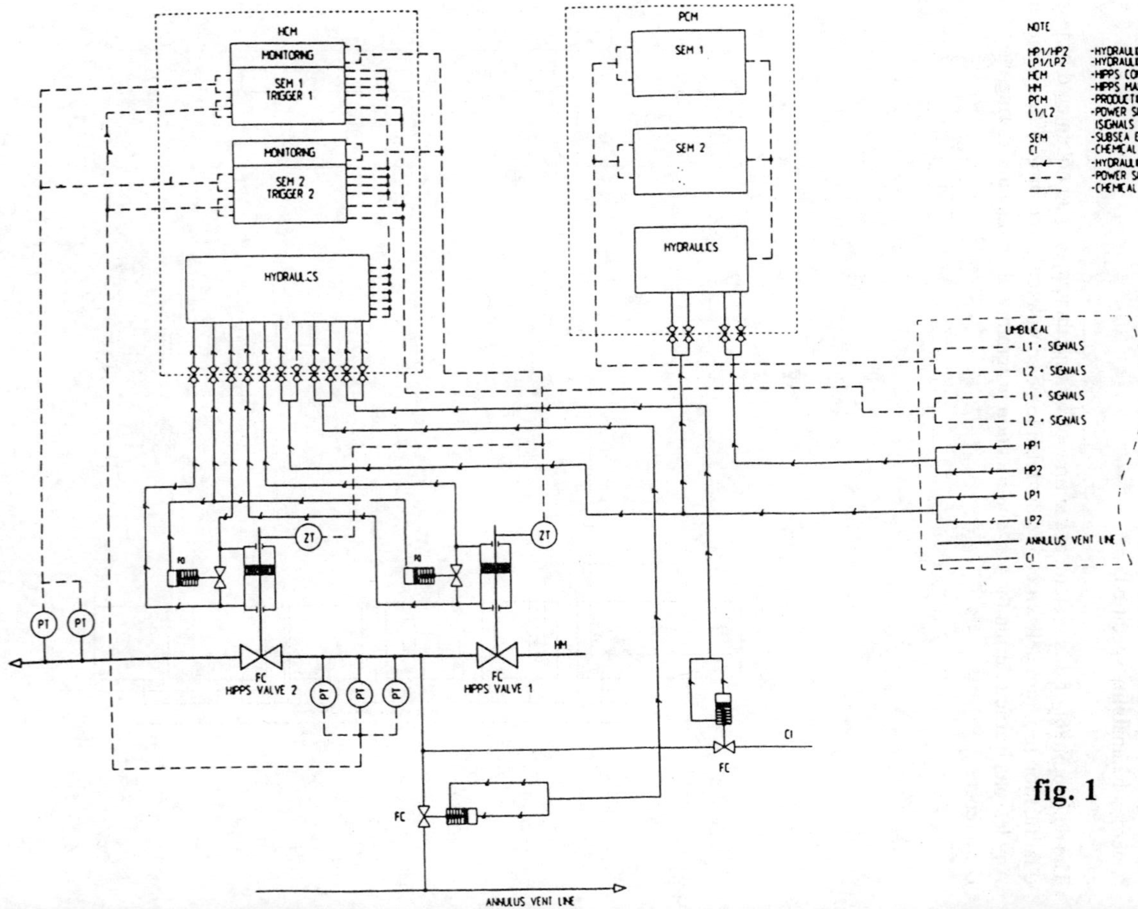
HCM
MONITORING
SEM 1
TRIGGER 1
MONITORING
SEM 2
TRIGGER 2
HYDRAULICS
PCM
SEM 1
SEM 2
HYDRAULICS
NOTE
HP1/HP2 -HYDRAULIC HIGH PRESSURE
LP1/LP2 -HYDRAULIC LOW PRESSURE
HCM -HPPS CONTROL MODULE
HM -HPPS MANIFOLD
PCM -PRODUCTION CONTROL MODULE
L1/L2 -POWER SUPPLY LINE (SIGNALS SUPER IMPOSED)
SEM -SUBSEA ELECTRONICS MODULE
CI -CHEMICAL INJECTION
-HYDRAULICS CONNECTION
-POWER SUPPLY LINE
-CHEMICAL INJECTION LINE
UMBILICAL
L1 + SIGNALS
L2 + SIGNALS
L1 + SIGNALS
L2 + SIGNALS
HP1
HP2
LP1
LP2
ANNULUS VENT LINE
CI
ZT
ZT
PT
PT
PT
PT
PT
FC
HPPS VALVE 2
FC
HPPS VALVE 1
HM
CI
FC
FC
ANNULUS VENT LINE

fig. 1

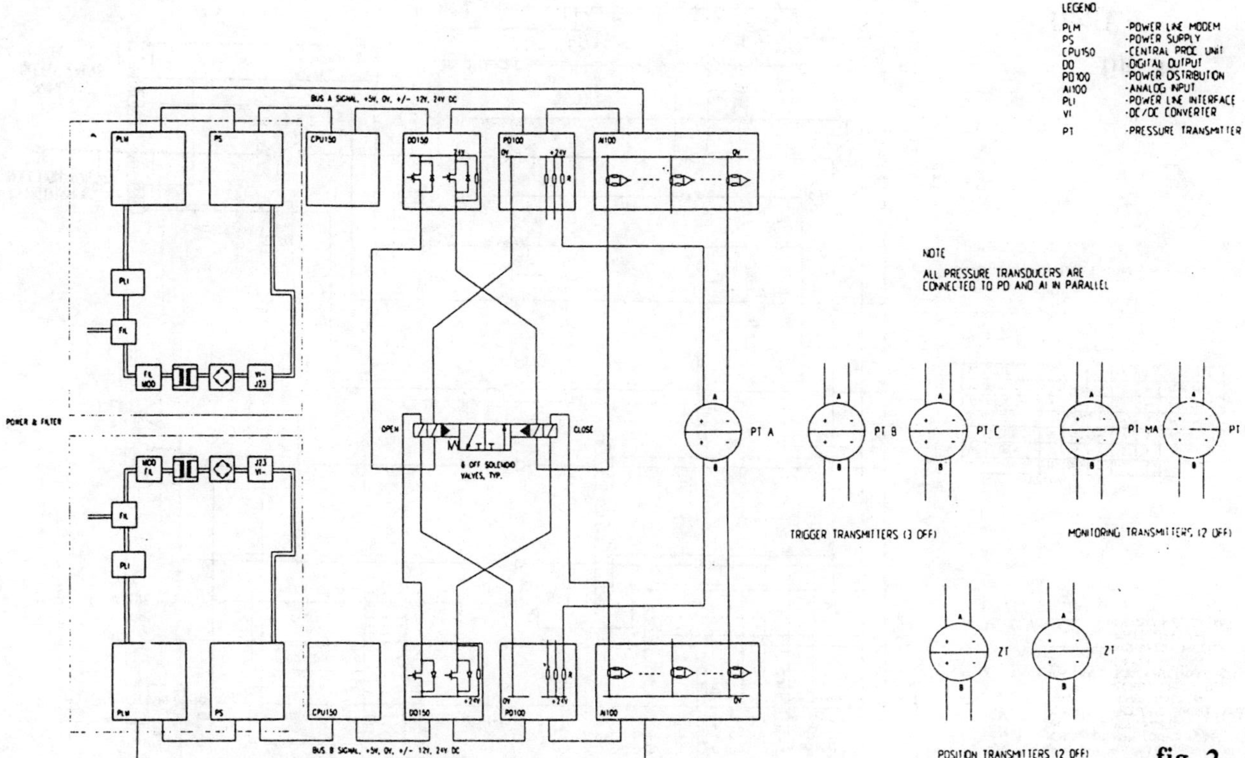
LEGEND
PLM -POWER LNE MODEM
PS -POWER SUPPLY
CPU150 -CENTRAL PROC UNIT
DO -DIGITAL OUTPUT
PD100 -POWER DISTRIBUTION
AI100 -ANALOG INPUT
PLI -POWER LNE INTERFACE
VI -DC/DC CONVERTER
PT -PRESSURE TRANSMITTER
NOTE
ALL PRESSURE TRANSDUCERS ARE CONNECTED TO PD AND AI IN PARALLEL
BUS A SIGNAL, +5V, 0V, +/- 12V, 24V DC
BUS B SIGNAL, +5V, 0V, +/- 12V, 24V DC
POWER & FILTER
PLM
PS
CPU150
DO150
PD100
AI100
PLI
FIL
FIL MOD
VI-J23
OPEN
CLOSE
6 OFF SOLENOD VALVES, TYP.
PT A
PT B
PT C
PT MA
PT MB
ZT
TRIGGER TRANSMITTERS (3 OFF)
MONITORING TRANSMITTERS (2 OFF)
POSITION TRANSMITTERS (2 OFF)
fig. 2

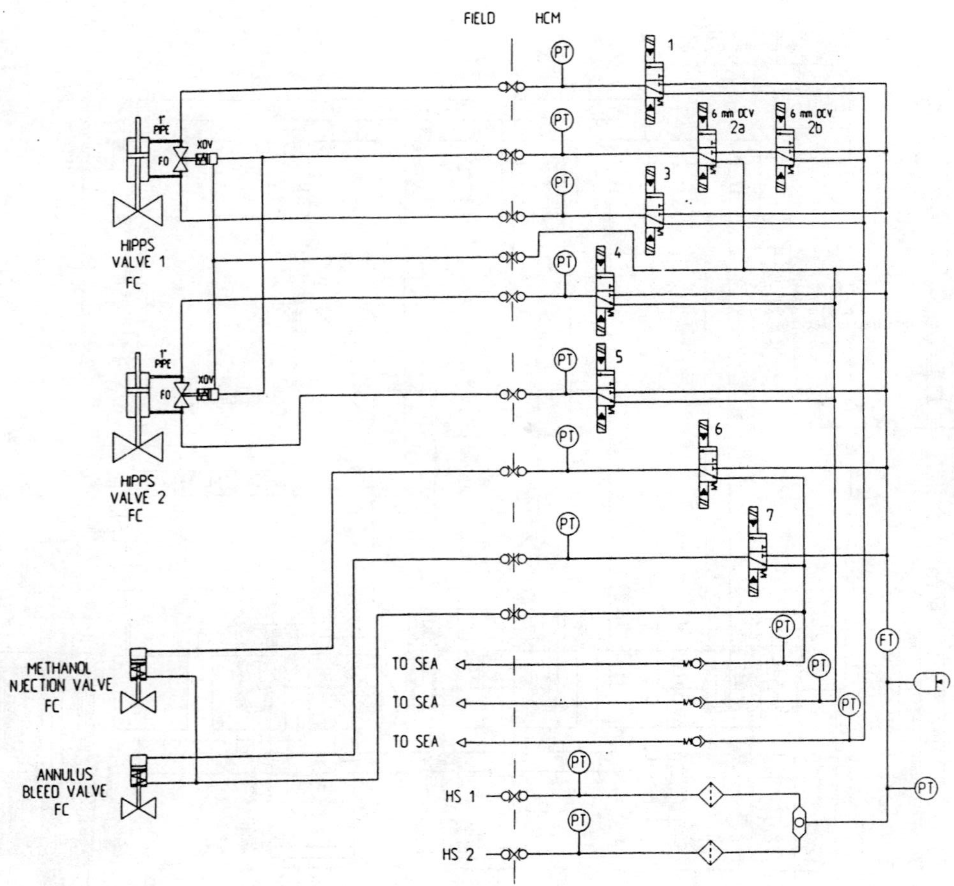


NOTES

PT -PRESSURE TRANSMITTER
FT -FLOW TRANSMITTER
XOV -XOVER VALVE
HCM -HIPPS CONTROL MODULE
DCV -DIRECTION CONTROL VALVE

ALL CONTROL LINES ARE 1/2" UNLESS OTHERWISE STATED

ALL DCV'S ARE "3 mm" UNLESS OTHERWISE STATED

LOGIC DIAGRAM

| | 1 | 2a/2b | 3 | 4 | 5 | |
|---|---|---|---|---|---|---|
| V1, V2 OPEN | S | S | V | S | V | AUTO MODE |
| V1, V2 CLOSED | V | V | V | V | V | AUTO MODE |
| V1 OPEN | S | S | V | - | - | MAN MODE |
| V1 CLOSED | V | S | S | - | - | MAN MODE |
| V2 OPEN | - | S | - | S | V | MAN MODE |
| V2 CLOSED | - | S | - | V | S | MAN MODE |

S = SUPPLY
V = VENT

**fig. 3**

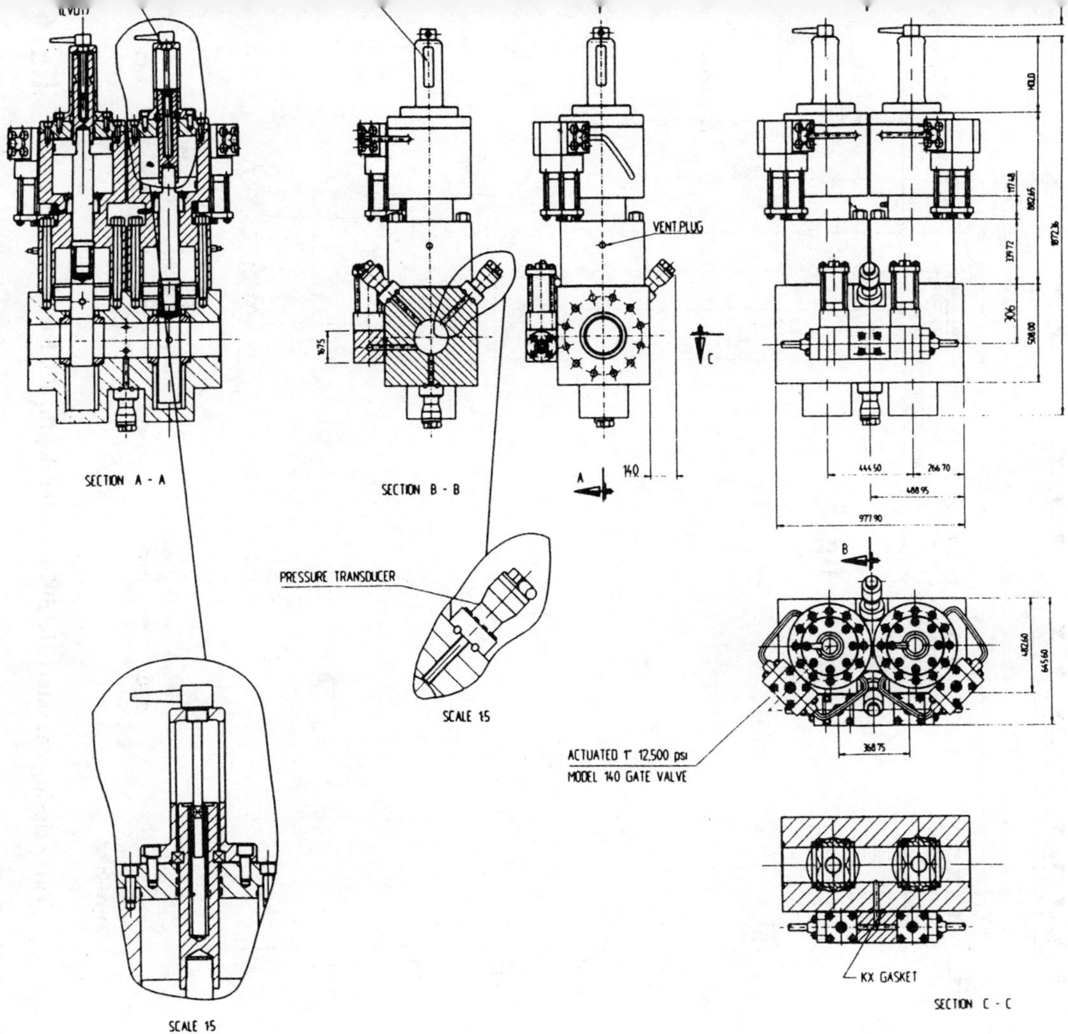


NOTES:

1 THIS DRAWING SERVES AS AN INDICATION OF THE DESIGN INTENTIONS ONLY

2 DIMENSIONS ARE APPROXIMATE ONLY

3 THE 1" XOVER VALVES ARE TO BE SUPPORTED FROM THE ACTUATORS

4 THE METHANOL INJECTION VALVE AND BLEED VALVE ARE TO BE PHYSICALLY CONNECTED TO THE VALVE BLOCK VIA A CONNECTION BLOCK

5 THE PROCESS CONNECTION BORE IN THE VALVE BLOCK FOR THE BLEED AND INJECTION VALVE IS TO BE 3/4"

6 THE PRESSURE SENSORS CONNECT TO THE VALVE BLOCK VIA KX-GASKETS

7 METHOD OF XOVER VALVE SUPPORT AND 1" INTERPORT CONNECTIONS TO BE DECIDED BY FMC

8 CONNECTION POINTS FOR 1/2" CONTROL LINE CONNECTIONS TO BE DECIDED BY FMC

9 PROCESS BORES FOR PRESSURE TRANSMITTERS ≤ 40MM

**fig. 4**

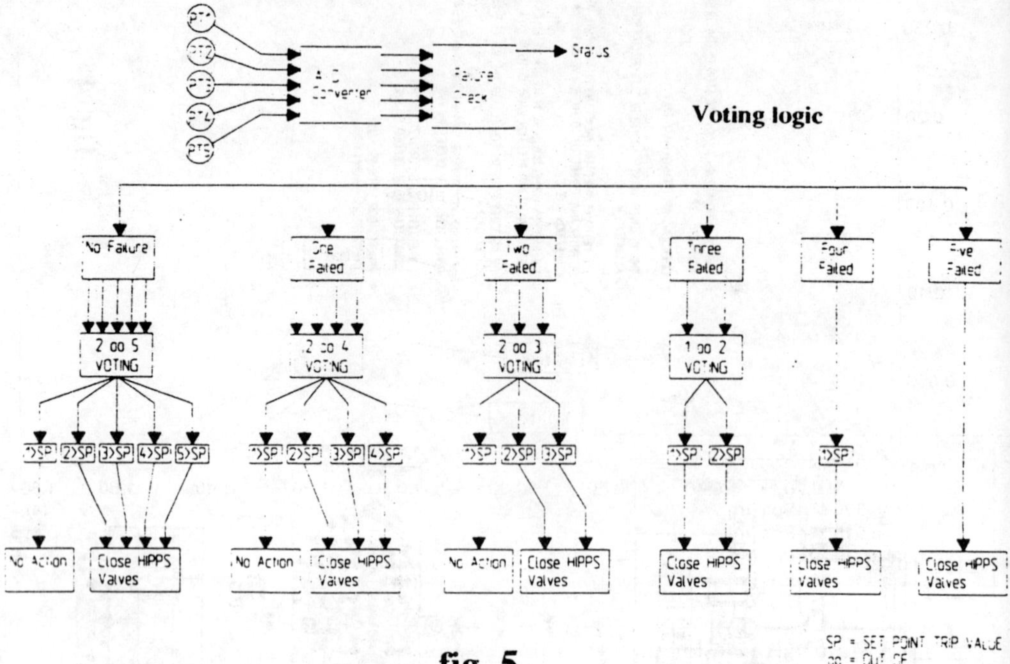


**fig. 5**

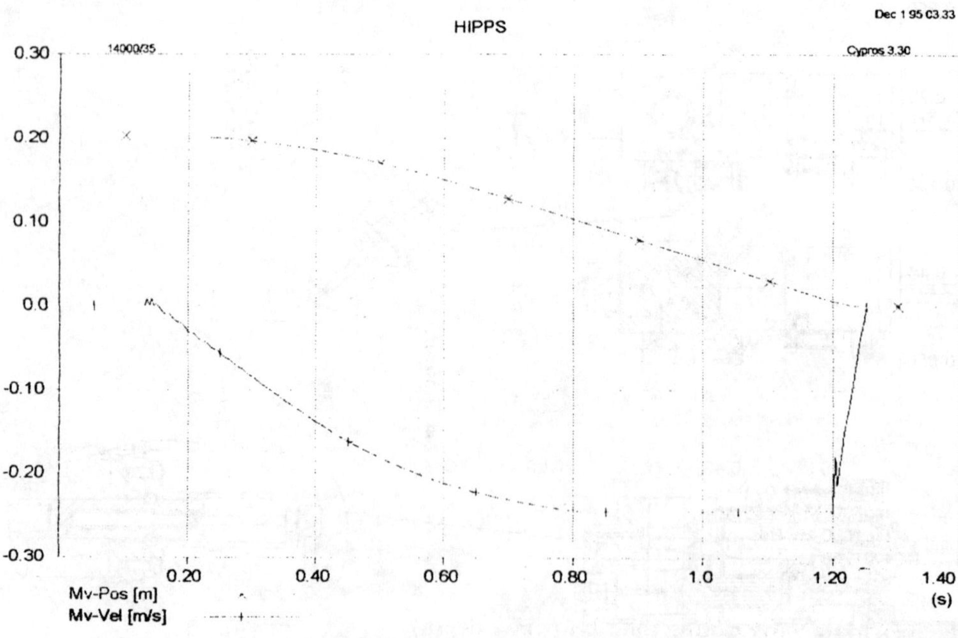


**Valve closing charateristic/ 500 m water depth, 170 bar process pressure** **fig. 6**

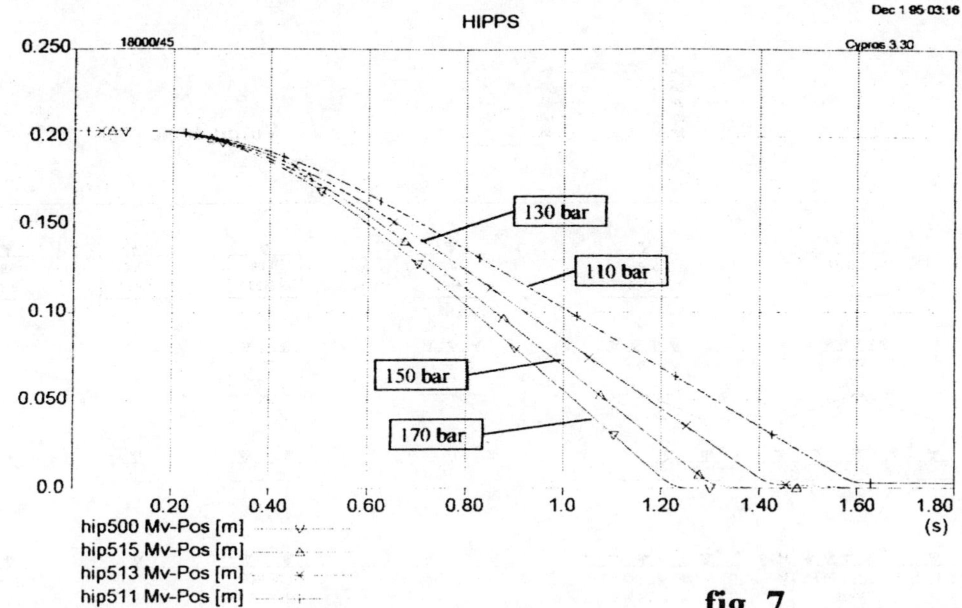


**fig. 7**

**Main valve closing time as f (process pressure)**

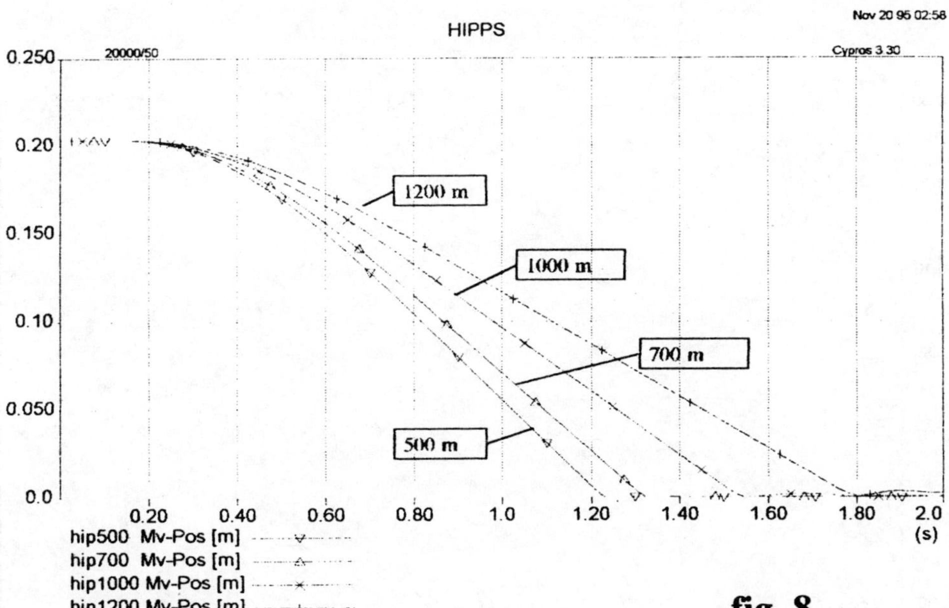


**fig. 8**

**Main valve closing time as f (water depth)**

# Extending the Technology

# Subsea power: electrical equipment demonstration (SPEED)

C J PEACHEY and G T BAYNES
GEC Alsthom Engineering Systems Limited

**In mature offshore oilfields, many new discoveries and untapped reserves lie in smaller more isolated reservoirs where exploitation using conventional platform based techniques would be uneconomic. There is considerable interest in developing these fields as subsea satellites tied back to existing infrastructure. Tie back distances are generally considered to be economical up to about 50km, beyond which pipeline and umbilical costs may outweigh the costs of modern re-usable platforms.**

**Depending on the reservoir characteristics of the field, mechanical subsea boosting techniques may be required to provide sufficient lift energy or to improve operability by producing favourable pipeline flow regimes. Although down hole pumping is well developed, multiphase pumping (e.g. Nautilus) and subsea separation (e.g. VASPS) are still at a demonstrator stage. All these subsea booster techniques require power to be delivered to the seabed. Due to pressure drops in the umbilicals, hydraulic drivers and transmissions are only effective over short distances. Electrical drives and transmissions are therefore essential to deal with step outs up to 50km. A key requirement for any of these boosting techniques is the availability of subsea electrical power transmission, distribution and control equipment. It is generally agreed that the development of marinized electrical systems is lagging behind the development of boosting equipment. This project aims to provide the technology to support such equipment.**

## Background

During 1992/93, GEC ALSTHOM Engineering Systems Limited was sponsored by BP International Limited to investigate the implications of installing electrical power distribution equipment subsea. The study compared the various types of transformer, switchgear and power electronics, and also the possible containment concepts. It was concluded that a power distribution system to feed a number of loads up to a total capacity of 5MVA can be developed from the proprietary equipment already well established for land based systems.

Despite this, manufacturers have not yet developed their products for subsea use because:

a) Oil companies are still evaluating the various boosting options and there is as yet no clear consensus on the rating, environmental and operating requirements. It is therefore difficult for manufacturers to decide what type of equipment to develop.

b) The offshore oil industry is the only potential customer, and there is no evidence that this market will ever become significant when compared with utilities, traction or other land based applications.

c) The technical resources and production facilities of many electrical equipment manufacturers are not suited to the type of mechanical engineering involved in packaging for subsea use.

For these reasons, the best way forward is for a joint venture between several oil companies and manufacturers to develop subsea electrical power distribution equipment. It should use a generic approach, which can be applied to a range of boosting techniques and system configurations. This is the philosophy behind this project.

Generic Technical Requirements

It is not realistic to assess different types of electrical equipment without giving some consideration to the interactions which will arise when they are built into a system. The performance of transformers, control and protection switchgear, and power electronics need to be assessed both at operating conditions and at fault conditions which are realistic. For the purposes of this study, systems similar to that shown in Figure 1 have been assumed. They could comprise several fixed speed and variable speed electric motor driven loads up to 1000kVA fed from a distribution system with a total capacity of 5MVA.

Inevitably, many of the factors affecting the choice of equipment are within the remit of the client/operator. Some considerations depend on the particular requirements, step out distances, and system control philosophies. Others, such as the grouping of components and mounting arrangements, depend on practices within the offshore industries for installation, intervention and retrieval. In view of the difficulty of intervention, high reliability and low maintenance are of the utmost importance. In this regard, it can be assumed that options requiring the least development will pose the lowest risk.

A frequency converter is required e.g. to provide a variable speed capability for induction motor driven pumps. Power electronics packages using gate turn-off devices are well established for such purposes and have been very successfully used by GEC ALSTHOM in traction applications (e.g. in the Networker and Eurostar (TMST) drive systems). They have, in turn, been superseded by IGBT devices, and we have adopted these for the Demonstrator. They are by no means the lowest cost option, but they are most suited to subsea use because they are more compact, and have lower losses than the GTOs. Cooling and atmospheric pressure enclosure sizes assume great importance in this application.

Clearly, seawater must be excluded absolutely from the electrical components by means of leak tight enclosures. The main types of containment are:

a) Pressure compensated enclosures, which incorporate a flexible membrane or bellows to ensure that the internal pressure equals the external seawater depth pressure. It would be impractical to equalize pressure in this way by using gas or air fillings due to the large volume changes with depth.

b) Atmospheric pressure enclosures, which are constructed as pressure vessels capable of withstanding the maximum external depth pressure. No particular difficulties are foreseen in achieving this. Submarine hulls operate at similar pressures and are generally much larger.

The insulating and heat transfer properties of the medium within the enclosure affect the operation and reliability of the electrical equipment. Another important consideration is whether to install equipment in individual enclosures or group them together in larger containments. As a result, the environmental compatibility of different types of equipment must be assessed. This aspect also has a bearing on requirements for maintenance and the reliability of electrical interconnections.

<u>Demonstrator</u>

For further progress to be made towards acceptance of subsea electrical systems in production oil fields, a representative 'demonstrator' system must be constructed and operated in realistic conditions. The project is organised in 3 main stages:

STAGE 1: Equipment Definition
STAGE 2: Manufacture and Test of Demonstrator Compnents
STAGE 3: Onshore and Subsea Trials

Depending on the confidence achieved from Stage 1, it may be practical to adapt Stage 2 to justify its immediate use in a Production System, so that following Stage 3 the hardware could be transferred to an operational unit of an oil company.

Stage 1: Equipment Definition

Operators of oil processing equipment are continuously reviewing the way in which marginal/deep-water fields will be developed. For this reason a steering committee has been established so that end-users can confirm that the electrical system and its deployment remains representative of production requirements. The preferred types of equipment have been deliberately chosen to minimise technical risk. However, some uncertainties remain, mainly related to possible cooling and contamination problems arising from compactly enclosing the equipment in a sealed containment. The activities at this stage are designed to given confidence in proceeding to manufacture. It is also important to do some work on 'second choice' options to provide back-up solutions. The work packages are:-

a) Assessment of Oil Cooled Transformers in pressure compensated enclosures at ratings up to 5MW.

b) Assessment of Switchgear in gas filled atmospheric pressure enclosures, with particular emphasis on cross contamination and electromagnetic compatibility between equipments in a sealed containment.

c) Studies of Electronic Variable Speed Converter components in gas filled atmospheric pressure enclosures at ratings of 325kVA and 1.2MVA, including circuit design and cooling.

d) Survey, comparison and selection of available proprietary subsea connectors.

e) Analytic assessment of Controls and Diagnostics, system methodologies and communication equipment, with particular emphasis on electromagnetic compatibility with power conditioning equipment.

f) Assessment of all the above equipment by undertaking a 'Failure Modes and Effects Analysis' (FMEA) at the lowest level possible at the time of equipment definition, i.e., at component level wherever possible.

Stage 2: Manufacture

The objective of this stage is to design and manufacture the units required to drive the 250kW load to be used for testing (Stage 3).

The demonstrator system is made up of a number of units, as shown in Figure 2. The ratings are large enough to be representative, whilst minimising the capital and energy costs of the project. For added flexibility, the three units can be combined in different ways to suit the desired test site and load or simply to permit staggered timescales for producing the units.

a) Main Transformer Unit

Detail design/drawing, manufacture and works testing of a 1MVA 11kV/3300V package.

The transformer is oil cooled, and installed in a pressure compensated enclosure similar to that which would be used as a main land-system transformer (up to 5MVA). Although a rating of only 325kVA is required for the test programme, the additional cost for the 1MVA unit is justified because the transformer is thereby made more usable in a future production system.

b) Load Switching Unit

Detail design/drawing, manufacture and works testing of a busbar and switchgear assembly.

This comprises both air and vacuum isolated switchgear in a gas filled atmospheric pressure enclosure. Power can be switched between at least two loads to permit overload and short circuit faults to be simulated. Cross-contamination and electromagnetic compatibility can be demonstrated.

Connectors are incorporated such that the effectiveness of wet and dry-mateable connectors can be demonstrated by interchanging members carrying similar supplies, and confirming both interchangeability and reliablity.

An additional 40kVA transformer is included to allow two typical auxiliary system loads to form part of the test programme e.g. to demonstrate valve mechanism drives.

c) Variable Speed Unit

Detail design/circuit layout, manufacture and works testing of a 325kVA sub-system.

This comprises a compensated transformer, with a different compensating mechanism from the 1MVA unit, linked to a control system housed in a gas filled atmospheric pressure enclosure. The frequency converter, in this case produced by CEGELEC, a GEC related company, is made up of rectifier, inverter and filter assemblies whose electronic components operate at less than 1kV. The transformer reduces the input voltage to this level. A closed circuit liquid cooling system may ultimately be required for the power electronic devices. Options are under consideration. For obvious reasons, the aim must be one of passive cooling rather than by using a pump or fans.

d) Control Equipment

Remote operation and monitoring of the power equipment requires design and manufacture of signal transmission and processing equipment. An optical fibre link will be used. Various sensors will also be needed. A multiplex system (GEC ALSTHOM Telecode 80) will be adopted to avoid the use of multicore cable. Standard PLC equipment will provide input and output facilities. Most of the sending-end control electronics will be housed within the Load Switching Module containment.

Stage 3: Testing and Trials

The test programme has to convince oil company operating groups that the demonstrator provides a full justification for adoption of similar systems for full scale production. It is difficult to anticipate the operator's specific demands, but it is clear that the system must be shown to function successfully when submerged in seawater. The electrical power should ideally be used to drive a pump over the full torque and speed range. Other loads need to be available so that fault conditions can be set up, and to allow switchgear performance to be demonstrated.

Prior to complete system submerged trials, individual items of equipment will have been tested at the manufacturing stage. Their enclosures will have been leak and pressure tested to full design depth (300m). The main transformer will be subjected to full current and maximum voltage while at this depth. This external pressure testing of special interfaces (e.g. connector seals) has been included in Stage 1.

a) Test Arrangement & Equipment

As indicated above, the dynamic load can be provided by an induction motor driven pump, which can be used to circulate water at appropriate heads and flows. The alternative of re-generation to reduce energy consumption would not provide sufficient flexibility for all the test regimes likely to be covered.

Pumps will be used for all loads in the submerged trials.

Flexible submersible cables have to be of sufficient length for connection of the supply to the transformer (see Figure 2). The loads will be shore based, so similar lengths of low voltage cable will be needed to connect these to the demonstrator modules. The 11kV supply is readily available from the existing power distribution systems, via associated switchgear.

Deployment for submerged trials will require installation procedures similar to those employed for subsea wellhead equipment, and the skids are designed to fit the smallest moonpool of a deployment vessel. This adds to the realism of the demonstrator. Otherwise, the installations are generally similar to those employed for land based distribution equipment. The receiving end of the diagnostic and control equipment, with data handling facilities, will be housed and manned as required during the full test programme.

b) Procedure & Techniques

A large part of the test programme duration will be taken up with an endurance trial. It is assumed that a major requirement is for reliable continuous operation at near steady conditions. This will be represented by running the demonstrator in seawater for a minimum total of 3000 hours at full load. A representative number of start/stops will be included, and the plant will be left idle for set periods to monitor the effect on re-starting. A range of operating parameters will be monitored, mostly to verify heat exchange reliability.

In addition to start/stops, more demanding transient tests will be carried out both before and after the endurance trial. Transportation to site, installation and retrieval activities will all involve shock and vibration, and will require a number of make/breaks for the wet-mateable connectors.

Load switching trials will be undertaken according to a pattern based on type-approval tests for the equipment as supplied for land-based applications. The variable frequency power electronics will be required to drive the dynamic load for several extended periods when performance over the full range of speed and torque is demanded. These tests can be based on those applied to the equipment in its industrial application.

Protection against faults, similar to that in land based practice, is incorporated. Faults will be simulated, as would be the case for a shore-based installation. These will include overloads up to the point at which the circuit is tripped out. Parameters such as temperatures at critical points will be monitored during such tests.
Performance of the protection system under conditions such as electrical penetrator failure will be demonstrated by introducing deliberate short circuit links. Long term water vapour permeation through seals, and the subsequent condensation, will be monitored by the diagnostic system. This will be proved before the endurance trial.

c) Test prior to Submergence

The demonstrator modules assembled as the electrical system shown in Figure 2 will be run in a GEC ALSTHOM test bed to confirm the correct operation of the inter-connected sub-systems. Sufficient continuous running will take place to verify the predicted rates of temperature rise at the locations monitored by the diagnostic system.

d) Submerged Trials

The current plan is to conduct the submerged trials in a dock at VSEL in Barrow. This gives the advantage of support from a sister company in water of sufficient depth to be realistic. There are established electrical supplies and back-up facilities to give the trials a high chance of first time success.
Deeper trials would be an extension to the project, noting that the components will have been subjected to full submerged pressure in the earlier stages.

## Technical & Economic Evaluation

On completion of the test programme, the equipment will be dismantled. Key interfaces such as terminations and heat sink thermal contact faces will be visually examined. Items such as fasteners, springs, bellows etc. will be assessed with reference to the reliability analysis. The observations will be included in the overall performance report on the demonstration project.

Towards the completion of the design and manufacturing phase of the project, data will have been accumulated on equipment selection criteria, procurement routes, cost, dimensions and reliability. This will provide an important contribution to the analysis of capital expenditure for subsea projects. Success in the testing phase will establish sufficient confidence to justify operators in proceeding to a production trial. The equipment demonstrated in this project will be available for such a trial, and its design depth (300m) is appropriate to applications where the alternative to sub-sea processing involves an expensive floating production system.

A preliminary economic assessment of the fundamental principle of subsea electric power distribution has already been carried out by examining the capital expenditure required for various likely system configurations. Figure 3 shows one possible production system operating at up to 21km from a platform. It was costed both for the components arranged as shown and also on the basis of providing individual cable links feeding each load. From the results of a range of such costings it was found that at distances greater than about

4km the "distribution option" became increasingly economically attractive compared with the "individual feed option".

More comprehensive estimates of capital and operating costs are emerging as the components for the demonstration system are fully defined within the SPEED project. As the project progresses, information will be generated on reliability, so that retrieval and maintenance schedules can be realistically postulated.

Design studies for applications in water depths of the order of 1000m can be based on the SPEED findings. The major variations will be in the construction of the pressure vessels. Wet mateable connectors are already proven for this depth.

<u>Project participants</u>

The SPEED project is being funded by BP, Shell, Texaco, Amerada Hess and GEC ALSTHOM. A grant is also being made available by the Offshore Oil and Gas Supplies Office of the UK Department of Trade and Industry. The 3 year programme is anticipated to complete in late-1997, and current indications are that it will do so.

GEC ALSTHOM

ENGINEERING SYSTEMS

# SUBSEA ELECTRICAL POWER DISTRIBUTION SYSTEM

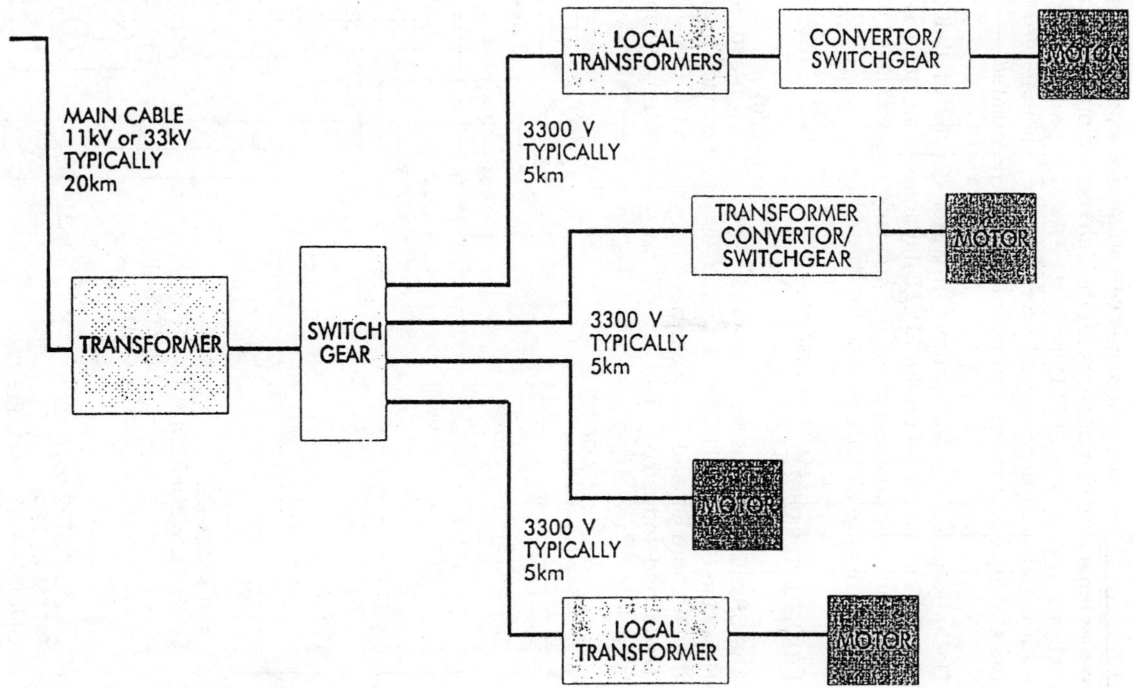


FIGURE 1

FIGURE 2



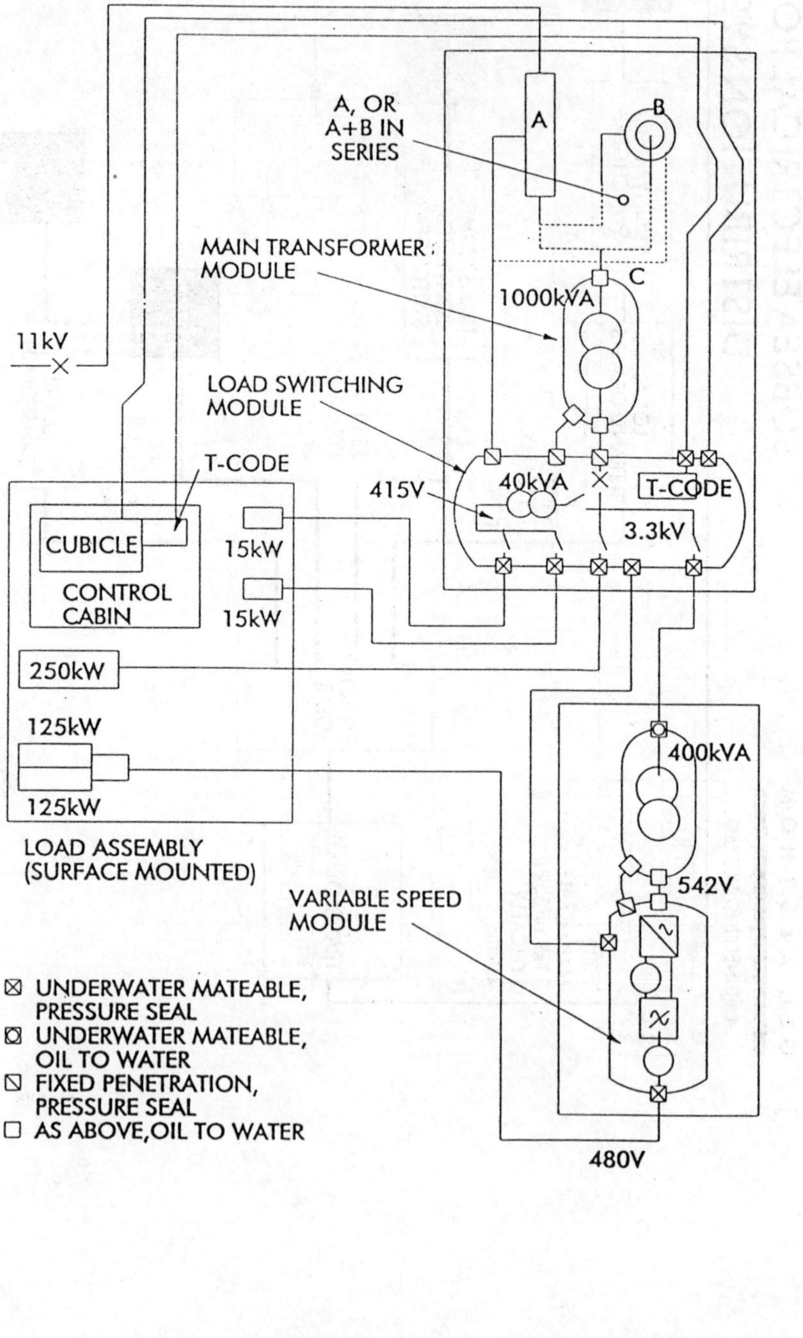
A, OR
A+B IN
SERIES
A
B
MAIN TRANSFORMER
MODULE
C
1000kVA
11kV
LOAD SWITCHING
MODULE
T-CODE
415V
40kVA
T-CODE
3.3kV
CUBICLE
CONTROL
CABIN
15kW
15kW
250kW
125kW
125kW
LOAD ASSEMBLY
(SURFACE MOUNTED)
400kVA
542V
VARIABLE SPEED
MODULE
480V
UNDERWATER MATEABLE, PRESSURE SEAL
UNDERWATER MATEABLE, OIL TO WATER
FIXED PENETRATION, PRESSURE SEAL
AS ABOVE, OIL TO WATER

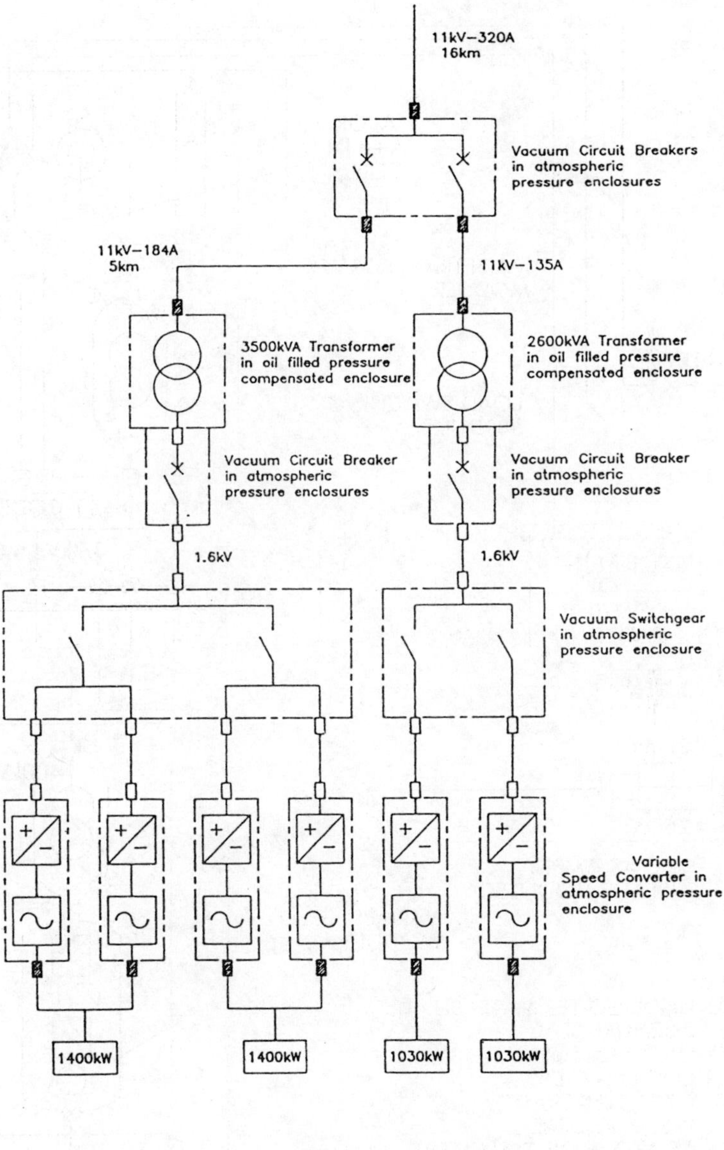
11kV–320A
16km
Vacuum Circuit Breakers in atmospheric pressure enclosures
11kV–184A
5km
11kV–135A
3500kVA Transformer in oil filled pressure compensated enclosure
2600kVA Transformer in oil filled pressure compensated enclosure
Vacuum Circuit Breaker in atmospheric pressure enclosures
Vacuum Circuit Breaker in atmospheric pressure enclosures
1.6kV
1.6kV
Vacuum Switchgear in atmospheric pressure enclosure
Variable Speed Converter in atmospheric pressure enclosure
1400kW
1400kW
1030kW
1030kW

FIGURE 3

# Jpdate on dynamic umbilicals

T HOLEN
Icatel Kabel Norge as

## 1.0 ABSTRACT

Ongoing and coming subsea field developments has increase in water depth, in the use of satellite well developments, and in the use of floating production systems.

Metallic umbilicals has several advantages over the traditional hose umbilicals with respect to permeation, fluid compatibility, hydraulic and mechanical properties.

The introduction of the metallic umbilical for dynamic applications is a cost effective solution for better dynamic umbilicals for future developments.

## 2.0 INTRODUCTION

Umbilicals, in general, are vital parts of the underwater production technology. But their reliability have been questioned, as many problems have occurred during laying and operation.

This is documented in "Study of the Performance and Reliability of Hydraulic, Electro-hydraulic and Multi-Functional Umbilicals" /1/. According to this report, improvements has to be performed in order to increase the reliability of the product. It is also relevant to compare the umbilical reliability with ordinary subsea cable reliability. The latter product group has a remarkable good track record, and "mature" umbilical designs should be expected to have the same kind of reliability.

One of the problem areas was that the inner liner of the thermoplastic hoses showed compatibility problems with some hydraulic fluids and methanol in special. To comply with

One of the problem areas was that the inner liner of the thermoplastic hoses showed compatibility problems with some hydraulic fluids and methanol in special. To comply with these compatibility problems the metal tube umbilical design was developed and installed, first for the static applications.

The following summarises what are considered to be the main advantages of metal tubes as hydraulic lines compared to thermoplastic hoses:

i) Fluid (methanol) permeation has been eliminated.

ii) Compatibility problems have been reduced

iii) Collapse resistance is drastically increased.

iv) Favourable response characteristics, especially for long length umbilicals.

v) No mechanical effect on other umbilical elements when pressurising.

vi) Higher weight/diameter ratio gives improved dynamic and seabed stability.

vii) Maximum umbilical manufacturing length without factory jointing is increased, as tube strings are jointed by welding in the lay-up process.

Dynamic umbilicals can be installed in several configurations; between a floater and the seabed, between one floater and a rigid structure, between two floaters or from a structure deck to the sea bed. What is common is that in such installations, the umbilicals are exposed to environmental loads and loads from the floater movements.

Up to now dynamic umbilicals have been equivalent to umbilical designs comprising thermoplastic hoses as hydraulic lines. With the problem areas above in mind we started the development of the dynamic metal tube umbilical.

The metallic tube umbilical has already proved itself in several static offshore applications.

This paper describes the main steps in the working method and development towards metallic tube umbilicals for dynamic applications.

Following our introduction of this new generation of umbilicals to the market, a steel tube umbilical for dynamic application installed on a TLP has been analysed and tested. This first installation of a dynamic steel tube umbilical will be performed within the summer 1996. The analysis and test results shows good margins to the requirement for service life time.

## 3.0 PARAMETERS AND EFFECTS GIVING IMPACT ON UMBILICAL DESIGN AND ANALYSIS

### 3.1 DESIGN ANALYSIS

The dynamic umbilical utilises the basic philosophy of stranding far more than the static umbilical. In the dynamic application all umbilical elements are sliding on their neighbours all the time. The rate of relative movement is proportional to the amount of global bend radius variation. From this, internal wear and material fatigue becomes central effects to control through the choice of materials and the umbilical design.

Experience from previous dynamic cable testing and installations are also utilised as far as the behaviour of cable elements and material properties are applicable.

The design work of a dynamic umbilical is an interactive process starting with a proposed cross section serving the control system need for electric, hydraulic and fibre optic conductors and the need for weight and strength guided by the product application.

Umbilical parameters, bend radii allowances in correspondence with simultaneously tensions, stiffness and global geometrical distances are parameters to be set before entering the work on determining the static configuration through static analysis. One has to trim the umbilical length to ensure that too extreme tensions and bending radii are avoided as the floater is in its extreme far, near and transverse positions.

When the static configuration and corresponding loads are found acceptable, dynamic analysis is to be performed. Such analysis simulates the umbilical in the installed configuration and give loads, curvatures and dynamic responses of the system as results.

After examining the dynamic analysis results for extremes one have to check for not acceptable stresses in the umbilical elements in these extreme cases.

Figure 1 shows as an example the cross section of one of the dynamically tested umbilicals.

### 3.2 SECTIONS OF SPECIAL INTEREST ALONG THE UMBILICAL PATH

The highest axial tensions are usually found in the hang off region. Accessories transferring these loads to the structure can be designed as interface dimensions are given and the expected loads are read out of the global dynamic analysis results.

In our opinion it is advantageous with respect to stress distribution to have some distance between hang off and the region of umbilical bending i. e. I-tube or turret outlet. Then the stress in the hang off region can be treated as pure axial stress in the umbilical.

One of the most critical regions with respect to fatigue loads is the area where the umbilical leaves the floater or vessel and enters the catenary configuration. I. E. from a region where the umbilical is supported by a structure with very high bending stiffness to a riser région where only the umbilical bending stiffness rules. In such regions it is common to use bending stiffeners / bending strain relievers (BSR's). The basics for these items is to restrict the umbilical bend radius by giving distributed change in bending stiffness from the rigid structure to the umbilical bending stiffness itself. A bending stiffener is designed to give a minimum bend radius at a certain umbilical axial tension.

Evaluations of buoys entrance and exits with respect to bend radii and fatigue loads has to be performed in a similar way.

In the umbilical touch down area, lateral movement has to be investigated with respect to possible abrasion between umbilical and seabed.

Tie-back and sea-bed anchorage accessories are also items to be checked for stress and fatigue loads.

## 3.3 FATIGUE EVALUATION METHOD DESCRIPTION

The scatter diagram, giving the number of waves for significant wave heights and wave periods are divided into 3 - 5 classes of wave heights. Fatigue dynamic analysis are then run per wave height class.

With basis in results from these analysis and the sum of wave cycles in each class, fatigue life assessment is performed using the deterministic method known as the Miner-Palmgren method.

These calculation results forms basis for fatigue evaluation and testing addressing the umbilical in the analysed scenario. Flex and fatigue testing programs is chosen according to the ruling philosophy and fatigue dynamic analysis results.

## 4.0 TESTING OF DYNAMIC METAL TUBE UMBILICALS

Dynamic umbilicals, in general, have to be tested for investigation of fatigue and internal wear properties.

With respect to fatigue testing of dynamic umbilicals one of two philosophies can be followed.

In both cases the actual I-tube or turret outlet situation is simulated. The test sample is equipped with the actual bend stiffener and a supporting structure model, and is exposed to axial tension and deflection angles of same magnitude as expected in the practical application. Number of bend cycles is chosen in accordance with the deflection angle i.e. sea-state and test schedule. Table 1 shows an example of a test parameter set.

The other way of testing is to build up a test program giving the same amount of fatigue damage as the calculated fatigue damage from the fatigue analysis.

## 4.1 FLEX TESTING

For the purpose of performing flexing test on complete umbilical samples, we have developed and built test rigs that can bend a tensioned test sample. The test rig principle is shown in figure 2.

The sample is tensioned to determined axial loads. The basis for the test parameters will in most cases be the results from dynamic analysis. From this we find the maximum tension and the variation in bending radius due to environmental impact such as wave and current. The applied tension in the rig is adjustable as a function of the bending angle.

Bending radius shall be restricted by a bends stiffener or sheave segments. The bend radius shall be in accordance with the allowed umbilical minimum bending radius at the actual tension.

The tubes are pressurised to working pressure throughout the test. Electrical and fibre optic elements are tested before and after the flex testing and are looped to ease continuity measurements during the tests.

The tests runs until the predetermined total number of cycles is achieved in each parameter class or until a failure occurs.

After the flexing test, pressure tests and electrical tests are performed in order to verify the quality of the tubes and electric cable.

In addition to testing on development samples, the flexing tests are performed on prototypes to verify that the umbilical and the bend stiffener withstands the dynamic responses during operation.
I.e. this flexing test is used as a qualification test for specific umbilical projects in which the loads and the test sequence will be designed according to the actual floating unit dynamics and environmental conditions.

## 4.2 FATIGUE TESTS ON TUBES

In addition to testing on complete umbilical samples, we perform fatigue tests on single tubes by the rotative bend test method. This is done to gather fatigue data on tubes, with and without welds, being in the condition they are as they go into the umbilicals.

## 4.3 OTHER TEST FACILITIES

Other test facilities used for mechanical testing of dynamic umbilicals are:

- Bending tensile test rig. Simulation of laying and retrieval.
- Figure of 8 test rig. Reverse bending test, previously used for fatigue testing.

The bending tensile test is basically a test which is designed for simulating the laying and retrieval operations. For this purpose there exists a recommendation for testing of large subsea power cables, see ref. /2/, where the cable shall be bent around a sheave, with the same radius as the actual laying wheel, and tensioned to maximum allowable tension. While under this tension, the wheel is rotated back and forth three times (or more). This mechanical test is followed by an electric test program in order to verify the quality of the cable. This philosophy can also be adopted for umbilicals. The mechanical test will be followed by an electric/hydraulic test program for verification of the umbilical quality.

The figure of 8 test is basically a test for simulating bends to which the umbilical will be subjected to during reeling and unreeling processes or other movements where the umbilical is not exposed to axial tension.

## 5.0 RESULTS

## 5.1 GENERAL EXPERIENCE WITH DYNAMIC DESIGNS

### 5.1.1 DYNAMIC POWER CABLES

Testing and installation of dynamic power cables, ROV cables and dynamic thermoplastic hose umbilicals has given information also utilised for the metal tube dynamic umbilical development.

The design philosophy, with respect to umbilical armouring package, sheaths and the internal designs of signal parts will be very similar to what is used in dynamic cable design. Experience with insulated power cores are obtained from installed dynamic power cables, of which one still is in service:

The Veslefrikk dynamic power and signal cable (12kV, 3x300 mm²). Two lengths are installed in a deep catenary configuration between Veslefrikk A (fixed platform) and Veslefrikk B (floater). The cables has been in service since the installation in 1989. No failure has been reported up to now.

The North East Frigg power and signal cable (12 kV, 3x25 mm²). At the NEF end of the 20 km route, the cable was installed in a small free hanging catenary from the seabed pontoon to the articulated column at the Field Control Station. This cable was installed for the predetermined ten years, and was taken out of service in 1993 as the field was abandoned. No failure was reported regarding the dynamic section. Previous to installation, samples of this cable was tested for up to 27 mill. cycles of bending during a test program 1981-82.

### 5.1.2 DYNAMIC THERMOPLASTIC HOSE UMBILICALS

For use at the Heidrun TLP we have performed design and dynamic analysis, produced, tested and delivered a dynamic umbilical comprising thermoplastic hoses and a relatively large combined power and signal cable element.

The Foinaven (West of Shetland) dynamic umbilicals was also designed on the basis of weight to diameter ratio; the umbilicals to have the same dynamic response as their neighbouring flexible risers. Analysis, and simulation testing is performed.

## 5.2 RESULTS FROM TESTS ON DYNAMIC METAL TUBE UMBILICALS

In our development work on dynamic metal tube umbilicals, we have followed two paths, one using titanium tubes and the other using stainless steel tubes as hydraulic lines.

From our tests, performed on several prototypes of metal tube umbilicals in our flex test rigs, we would present:

### 5.2.1 UMBILICALS WITH TITANIUM TUBES

Electro-hydraulic umbilical with 1/2" titanium tubes were flex tested for 3 000 000 cycles and stopped. Two welds broke down by fatigue cracking in the heat affected zone at an early stage. No failure occurred in the remaining tubes nor in the electric cable.

Titanium tube umbilical with large centre tube and 1/2" tubes (design similar to figure 1.) was tested for 900 000 cycles before one tube and the centre tube broke down by fatigue cracking of plain tubes. No failure occurred in the other tubes or electrical parts.

### 5.2.2 UMBILICALS WITH STEEL TUBES

Steel tube umbilical with 2" centre tube and 5/8" tubes (ref. figure 1) were tested aiming at no breakdown before 1.5 mill cycles. The sample had a breakdown at 1.9 mill cycles. The failure was fatigue cracking of the centre tube. No failure in other tubes or electrical parts.

Testing of a steel tube design with 2.5" service line and 1/2" steel tubes has been completed. The sample reached 1.35 mill cycles before breakdown. The failure was fatigue cracking of the centre tube starting from a tooth mark in the outer surface. No failure in other tubes or electrical parts.

Post test strip down examination of all samples showed only expected and acceptable amounts of internal abrasion between the elements in a layer and between layers.

Common for these test, is that the test samples are exposed to 5 m radius sheave support and simultaneously 100 - 200 kN axial tension. In addition the hydraulic lines are pressurised to working pressure.

In total this is a severe test method gives severe fatigue strain in the test sample elements, especially the large centre tubes.

A prototype of steel tube umbilical and BSR designed for installation on a tension leg platform in 900 m water depth in the Gulf of Mexico has also been tested. This test was fulfilled and stopped at 1.35 mill cycles. The sample had then been exposed to loads from more than 100 years of service life. No breakdown occurred. Post test strip down showed only expected amount of internal abrasion.

## 5.3 FATIGUE TESTS ON SINGLE TUBES

This program is ongoing, and results are continuously register into S-N diagrams.

Tests are performed on 1/2'' and 1.2'' tubes. In some of the samples welds has been included.

All results, so far, are found above the recommended limit fatigue curves for steel structures given in Norwegian Standard NS3472 (similar to the DOEB-curve).

## 6.0 CONCLUSIONS

Our ongoing in-house development program has shown that metal tube umbilicals are well suited for dynamic application provided proper control of the bending radii and thereby the fatigue damage.

The metal tube designs shows proper resistance with respect to internal abrasion and fatigue.

## 7.0 NOMENCLATURE

TLP Tension Leg Platform
WP Working Pressure

## 8.0 REFERENCES

/1/ Study of the Performance and Reliability of Hydraulic, Electrohydraulic and Multifunctional Umbilicals" by P H Knight, S G O`Mahony & D C Ricketts. Report ERC(W)C.24.529, July 1990. Department of Energy, Palace St, London.

/2/ CIGRE: "Recommendations for mechanical testing of submarine cables", published in Electra no. 68.

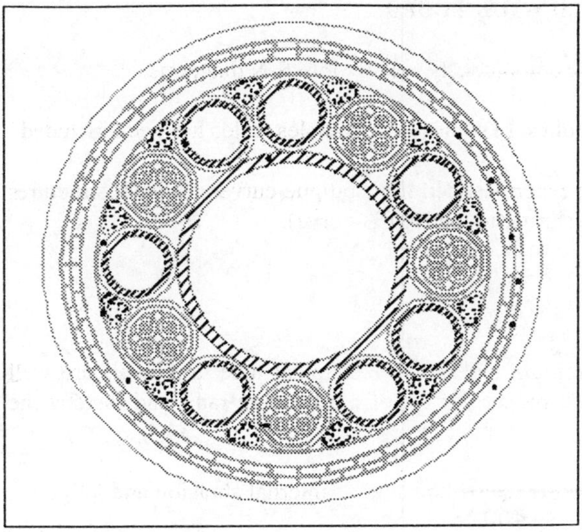

Fig. 1: Steel Tube Umbilical with integrated service line

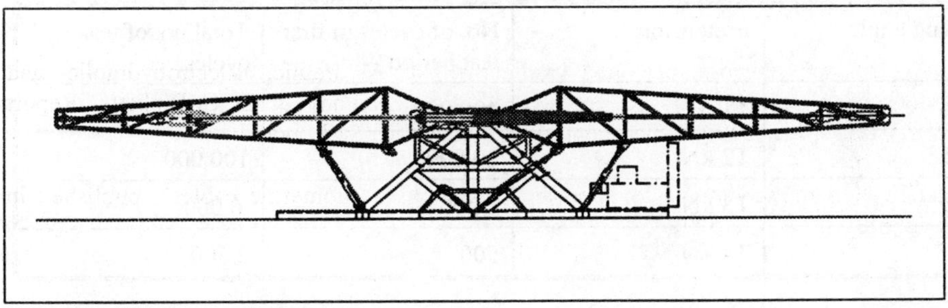

Fig.: 2 Flex rig test set-up

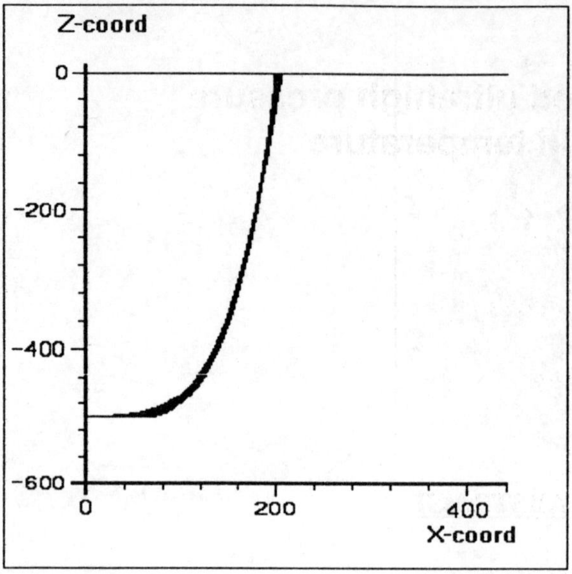


Fig.: 3 Motion envelope result plot

| Bending angle | Pretension | No. of cycles in first test period | Total no. of test cycles |
|---|---|---|---|
| +/- a° | T1 kN | 400 000 | 2 400 000 |
| +/- b° | T2 kN | 20 000 | 100 000 |
| +/- g° | T3 kN | 1 200 | 6 000 |
| +/- J° | T4 kN | 500 | 2 500 |

Table 1 Typical flexing test parameter set

# rototype fibre-optic based ultrahigh pressure emote sensor with built-in temperature ompensation

RAO and D A JACKSON
iversity of Kent

## ABSTRACT

A prototype single-mode fibre-based ultrahigh pressure remote sensor based upon a Fizeau cavity using a dual-wavelength coherence reading technique with built-in temperature compensation has been constructed capable of remote operation at distances up to 10km. A separate fibre-based temperature sensor with similar cavity length was incorporated into the pressure probe to allow the pressure measurement to be corrected for the temperature dependence of the pressure sensor. A range to resolution of ~$1.67X10^4$:1 and an overall measurement accuracy of ~±0.69% over a pressure range of ~1000bar have been achieved. This system represents a potential approach for the measurement of bore hole pressure in subsea control.

## 1. INTRODUCTION

For ultrahigh pressure measurements (up to 1000bar) in subsea control, extrinsic fibre-optic sensors based upon interferometry can offer many advantages over conventional electrical transmitters, such as inherent immunity to electromagnetic interference, safety in hazardous or explosive environments, high sensitivity, long-distance remote measurements, etc. A small extrinsic fibre-optic sensor based upon the deformation of a hollow glass microsphere in a miniature Fabry-Perot interferometer formed by the inner surfaces of the microsphere has been recently described, however the method of interrogation based upon the measurement of the change in wavelength of peak reflection restricts its operational range, giving a small range to resolution [1]. In this paper, we report, to our knowledge, the first prototype high resolution fibre-optic-based interferometric ultrahigh pressure sensor with built-in temperature compensation using coherence reading signal processing scheme and dual-wavelength technique, capable of truly remote operation. The cavity length change in the pressure probe, induced by thermal expansion, has been compensated by the means of incorporating a separate fibre-optic-based temperature sensor with similar cavity structure and optical path length into the pressure sensor and hence a high measurement accuracy was achieved. This prototype sensor could find applications in the measurement of high pressure in subsea control applications.

## 2. SENSOR DESIGN

The schematic diagram of the pressure probe (PP) head is shown in Fig. 1a. The sensing cavity is formed between the cleaved end of a 1.3μm single-mode fibre and the surface of a partially hollow cylinder used as the sensing element. The end surface of the sensing cylinder was coated with a thin Aluminium layer to form a mirror. The cavity length was set in the ~700μm region, which corresponds to the centre of the first valley of the autocorrelation function of the multimode laser diode (MLD) used as the low coherence source to illuminate the system [2]. As this gap is much larger than the diameter of the fibre core, only the first order reflected beam from the fibre end and a small part of the first order reflected beam from the cylinder surface will contribute to the interferometric signal. So the cavity can be properly termed an extrinsic Fizeau fibre interferometer [3], and the normalized output signal obtained is similar to that of a two-beam interferometer operating in reflection and is given by equation (1) when its output is processed with a remote Michelson interferometer [4]:

$$I/I_o = \frac{1}{2}\{[R_1+\alpha(1-R_1)^2R_2]+[(1-R_1)(\alpha R_1R_2)^{1/2}]\cos(\Phi_{P(\lambda n)}+\pi)\} \tag{1}$$

where $I_o$ is the total incident optical intensity. $R_1$ and $R_2$ are the intensity reflection coefficients of the fibre end and the diaphragm surface, respectively, $\alpha$ is the intensity coupling efficiency of the first beam reflected from the diaphragm surface. The optical phase change $\Phi_{P(\lambda n)}$ (where n is 1 or 2), corresponding to the displacement of the sensing cylinder at the mean wavelength $\lambda_n$ of either light source, takes the form:

$$\Phi_{P(\lambda n)} = \frac{2\pi}{\lambda n}(\Delta L_P - \Delta L_{LRI} + \Delta L_{PT}) \tag{2}$$

where $\Delta L_P$ is the optical path difference (OPD) change of the Fizeau cavity generated by pressure change applied to the sensing cylinder, and $\Delta L_{PT}$ is the undesirable OPD change due to the thermal expansion in the PP cavity and $\Delta L_{LRI}$ the OPD in the local receiving interferometer (LRI) which is a bulk Michelson interferometer. By measuring $\Phi_{P(\lambda 1)}$ or $\Phi_{P(\lambda 2)}$, we can obtain the absolute value of the phase change (modulo 2π) at each wavelength. The fringe number is determined by the difference of $\Phi_{P(\lambda 2)}$ and $\Phi_{P(\lambda 1)}$ which takes the form:

$$\Delta\Phi_P = \frac{2\pi}{\lambda e}(\Delta L_P + \Delta_{PT} - \Delta L_{LRI}) \quad (3)$$

where $\lambda e = (\lambda 1\ \lambda 2)/(\lambda 1 - \lambda 2)$ is the effective wavelength and is much larger than $\lambda n$. For the wavelengths of $\lambda 1$=1294nm and $\lambda 2$=1300nm used here, $\lambda e \approx 280\mu m$, which defines the unambiguous measurement range of the sensor. In this work, $R1 \approx 0.035$ and $R2 \approx 1$, hence expression (1) can be simplified as:

$$I/Io = \frac{1}{2}(R1+\alpha)[1 - \frac{(\alpha R1)^{1/2}}{(R1+\alpha)} \cos\Phi_{P(\lambda n)}] \quad (4)$$

The absolute value V of the visibility of the interferometric signal is:

$$V = \frac{(\alpha R1)^{1/2}}{(R1+\alpha)} \quad (5)$$

It can be seen from the expression (5) that when $R1=\alpha$, the maximum visibility can be obtained, which is equal to 0.5.

The optimum cavity length of the sensor was achieved by adjusting the relative positions between the cleaved end of the fibre and the surface of the sensing cylinder to obtain the maximum signal amplitude by using three screws with compression springs that connected the holder of the fibre and the support of the cylinder. When the OPD of the Fizeau cavity in the sensor head matched the OPD of the LRI, the holder of the fibre was bonded to the support of the cylinder by using adhesive. Rapid emergence to the optimum position was achieved by actively modulating the OPD of the LRI during this procedure. In a final version, the adjustment procedure would be similar except that the fibre could be mounted on a computer controlled mount with XYZ, roll and pitch adjustments such that the fibre could be correctly positioned. It would then be permanently bonded in place using electron or laser beam welding, giving very high mechanical stability. The manually adjustable arrangement based on the spring loaded screw shown in Fig. 1a is no longer required.

The sensing cavity of the temperature probe (TP) is formed between the cleaved end of the single-mode input fibre and the cleaved end of a short section multimode fibre. The end of the multimode fibre was coated with a thin Aluminium layer to form a mirror. The cavity length was also set in the region of ~700µm, which is identical with that of the PP. Therefore, the cavities of both PP and TP can be interrogated simultaneously by the matched LRI. The optical phase change, $\Phi_{T(\lambda n)}$, corresponding to the thermal expansion of the stainless steel tube at the mean wavelength $\lambda n$ of either laser which holds both of the fibres, takes the form:

$$\Phi_{T(\lambda n)} = \frac{2\pi}{\lambda n}(\Delta L_T - \Delta L_{LRI}) \quad (6)$$

where $\Delta L_T$ is the OPD change of the Fizeau cavity generated by the effective thermal expansion of the stainless steel tube given by:

$$\Delta L_T = 2[\alpha_s L - \alpha_f(L-d)] \tag{7}$$

where $\alpha_s$ and $\alpha_f$ are the linear thermal expansion coefficients of the stainless steel tube and the fibres respectively. The expansion of the stainless steel tube dominates the overall thermal expansion of the sensor as $\alpha_s >> \alpha_f$. Furthermore, as L>>d, the effective thermal expansion length is amplified mechanically in the ratio ~L:d, which will result in a significant improvement of the sensor sensitivity.

It is convenient to write,

$$\Delta L_{PT} = \beta \, \Delta L_T \tag{8}$$

where $\beta$ is a constant. By measuring $\Phi_{T(\lambda n)}$ or $\Delta L_T$, we can determine $\Delta L_{PT}$ since $\beta$ can be determined by experiment. By subtracting the phase change induced by $\Delta L_{PT}$ from $\Phi_{P(\lambda n)}$ given by equation (2), the thermally induced errors in the pressure measurement can be eliminated to first order. The ideal case is that when $\beta$=1 the temperature compensation can be achieved automatically.

The sensor head of the TP was assembled in the same way as the PP: the coated multimode fibre was first bonded to the inside of the stainless steel tube (outside diameter of ~1.6mm) and then the cap was carefully slipped over the end of the single-mode fibre; When the OPD of the Fizeau cavity in the sensor head matched the OPD of the LRI, the single-mode fibre was bonded to the stainless steel tube with adhesive. In a final version, both of the PP and TP would be permanently bonded in place using electron or laser beam welding, giving very high mechanical stability, as the curving and ageing effects of the adhesive could result in the very slow drift of the signals.

## 3. EXPERIMENTAL SYSTEM

The schematic diagram of the experimental system is shown in Fig. 1b. A serrodyne signal processing method [5] and a dual-wavelength coherence reading technique [6-7] are exploited to determine the optical phase change of the sensor. The low coherence sources used here were two multimode laser diodes (Mitsubishi 7701), with central lasing wavelength of 1294nm and 1300nm, respectively, and 5mW output power. The advantages of using a multimode laser diode as the low coherence source are its low cost and that more optical power can be coupled into single-mode fibres. This high coupling efficiency is due to the laser's small emitting area and N.A. which matches well with single-mode fibres, thereby resulting in significant improvements in the overall resolution, when compared with other low coherence sources [2]. The output of the multimode laser diode was first collimated with a 10X objective and then launched into the input port of a directional single-mode fibre coupler by another 10X objective. The undesirable reflection from the unused output port of the coupler was suppressed using index-matching gel. One of the mirrors (M2) in the LRI was mounted on a PZT element driven by a sawtooth voltage waveform with an amplitude of $\lambda_2/2$ (here $\lambda_2 > \lambda_1$) at frequency $w_c$ with a high voltage amplifier. The interferometric signal beam from the PP or TP was collimated by a 10X objective and then launched into the LRI and detected by the photodetector PD. The detected signal from the PP or TP, selected by changing the position of the fibre output port of the PP or TP with a translation stage (TS) and then focusing it on the PD, after bandpass filtering at this frequency was of the form of a heterodyne carrier at $w_c$ where its phase was modulated by the pressure or temperature induced phase changes in the interferometric signal. This signal was transfered to the lock-in amplifier. The sawtooth waveform was also bandpass filtered at $w_c$ to give a reference phase signal for the lock-in amplifier allowing $\Phi_{P(\lambda n)}$ or $\Phi_{T(\lambda n)}$ (or $\Delta L_P$ or $\Delta L_T$) to be determined. $\Phi_{P(\lambda 1)}$ and $\Phi_{P(\lambda 2)}$ (or $\Phi_{T(\lambda 1)}$ and $\Phi_{T(\lambda 2)}$) were determined by sequentially switching on laser diode 1 (MLD1) and laser diode 2 (MLD2). The lengths of the 1.3μm single-mode fibre links connecting the PP and TP with the signal processing box were ~2.9km and 3.3km respectively (as the optical fibre loss at this wavelength is so low (~0.36dB/km) deployment of the sensor can be up to 10km in practice). The pressure test system consisted of a hydrostatic pressure supply and a commercial precision pressure transmitter (KELLER-33) with a stated accuracy of ±0.2% of full scale. The temperature test system consisted of a precision commercial 4-wire platinum resistance temperature probe (Keithley) and a platinum resistance detector lineariser with an accuracy of ±0.2°C. The output phase signal from the PP or TP was sent to a personal computer through a 12bit analog-to-digital (A/D) converter to compare with the output of the pressure or temperature reference.

## 4. RESULTS AND CONCLUSION

The test results of the PP performance shown in Fig. 2a and b give the absolute phase change and the phase difference versus pressure, respectively. The total measurement range was ~1000bar, which produced a cylinder displacement of ~30μm. The mean optical phase response of the PP was ~0.09 rad/bar. The phase resolution of the lock-in amplifier was 1 degree so the displacement resolution was ~1.8nm, corresponding to a mean pressure resolution of ~0.06bar. Therefore, the achieved range to resolution was ~$1.67 \times 10^4$:1. The performance of the TP is shown in Fig. 3. The mean optical phase response of the TP versus temperature was measured as 0.56 rad/°C over a temperature range of ~17°C and the obtained resolution was ~0.01°C. The undesirable phase change in the PP induced by thermal expansion when the applied pressure was kept constant is shown in Fig. 4, and the mean optical phase response to temperature was ~0.68 rad/°C (in a final version this relatively high response coefficient can be reduced by at least a factor of 10 by bonding the fibre to the front face of its support). Hence, the value of β given by equation (8) was ~1.21. By subtracting the phase change resulting from the thermal expansion from the total phase output in the PP, we compensated for the error in the pressure measurement caused by thermal expansion. The results shown in Fig. 5 give a measurement uncertainty of ±0.15% over the full range. The repeatability, hysteresis, and nonlinearity of the PP over the full pressure range were ±0.2%, ±0.5%, and ±0.4%, respectively. Therefore, the overall measurement accuracy achieved was $\pm(0.15^2+0.2^2+0.5^2+0.4^2)^{1/2}(\%) \approx \pm 0.69\%$ over the full range. If the fibres in the PP and TP would be bonded using electron or laser beam welding (as previously discussed) the temperature range of operation could be increased to ~150°C. Over such a range the scale factor of the PP would vary considerably due to the change in the mechanical properties of the sensing cylinder, such as Young's modulus. However, since the temperature is measured to a high degree of accuracy in real-time, the measurement can be used to correct the pressure reading according to temperature tests. Furthermore in a final version one would need to stabilize the temperature of the light sources to achieve a long-term stability of 0.02nm to control the thermal drift of the central wavelength difference between both sources.

In conclusion, a prototype fibre-optic-based Fizeau interferometric ultrahigh pressure remote sensor using dual-wavelength coherence reading with built-in temperature compensation has been developed. A range to resolution of $\sim 1.67 \times 10^4$:1 and an accuracy of ±0.69% over a measurement range of ~1000bar have been achieved. By using advanced assembly techniques such as laser or electron beam welding in a final version, this prototype system could readily be developed as a practical approach for many ultrahigh-pressure sensing applications, particularly the measurement of bore hole pressure in subsea control applications.

ACKNOWLEDGEMENT

This work forms part of a collaborative project with Queensgate Instruments Ltd., supported by the SERC/DTI, U.K., under the LINK Nanotechnology Programme. The authors acknowledge Platon Ltd. (U.K.) for designing and fabricating the sensing cylinder.

## References

1. XU, M. G., and DAKIN, J. P.: 'A novel hollow-glass microsphere sensor for monitoring high hydrostatic pressure', Proc. SPIE, 1990, **1795**, pp. 2-7
2. GERGES, A. S., NEWSON, T. P., and JACKSON, D. A.: 'Coherence tuned fibre optic sensing system, with self-initialization, based on a multimode laser diode'. Appl. Opt., 1990, **29**, pp. 4473-4480
3. MURPHY, K. A., GUNTHER, M. F., VENGSARKAR, A. M., and CLAUS, R. O.: 'Quadrature phase-shifted, extrinsic Fabry-Perot optical fibre sensors'. Opt. Lett., 1991, **16**, pp. 273-275
4. SANTOS, J. L., LEITE, A. P., and D. A. JACKSON, D. A.: 'Optical fibre sensing

with a low-finesse Fabry-Perot cavity', Appl. Opt., 1992, **31**, PP. 7361-7366.
5. JACKSON, D. A., KERSEY, A. D., CORKE, M., and JONES, J. D. C.: 'Pseudo-heterodyne detection scheme for optical interferometers'. Electron. Lett., 1982, **18**, pp. 1081-1082
6. KERSEY, A. D., and DANDRIDGE, A.: 'Dual-wavelength approach to interferometric sensing'. Proc. SPIE, 1987, **798**, pp. 176-181
7. RAO, Y. J., and JACKSON, D. A.: 'Prototype fibre-optic-based pressure probe with built-in temperature compensation with signal recovery by coherence reading', Appl. Opt., 1993, **32**, pp. 7110-7113.

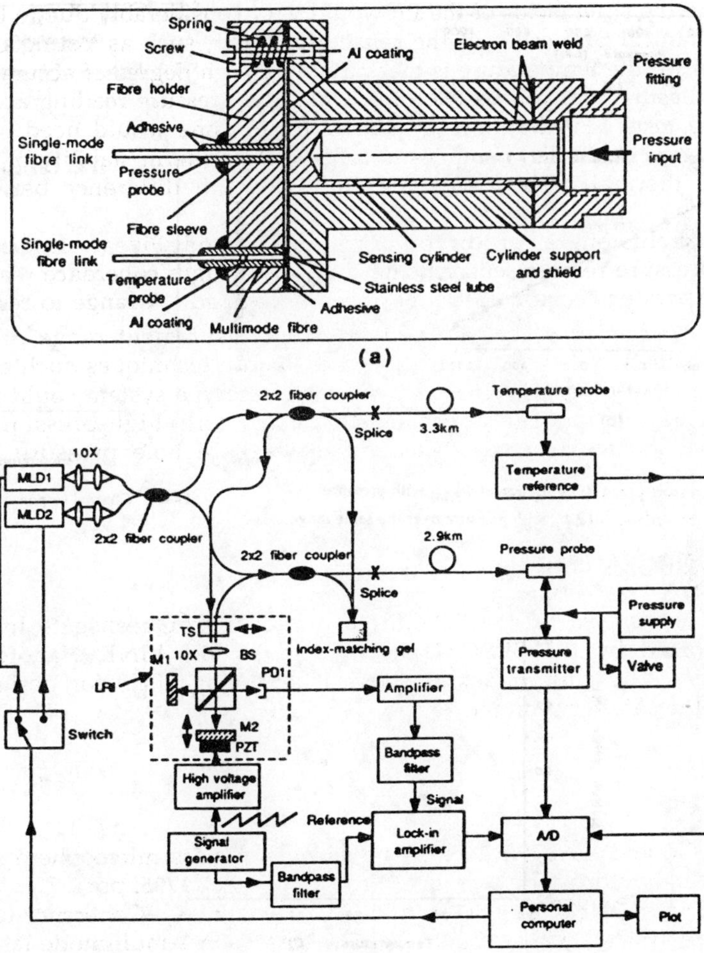


Fig. 1 Schematic diagram of the sensor and experimental set-up
a Sensor head
b Experimental system

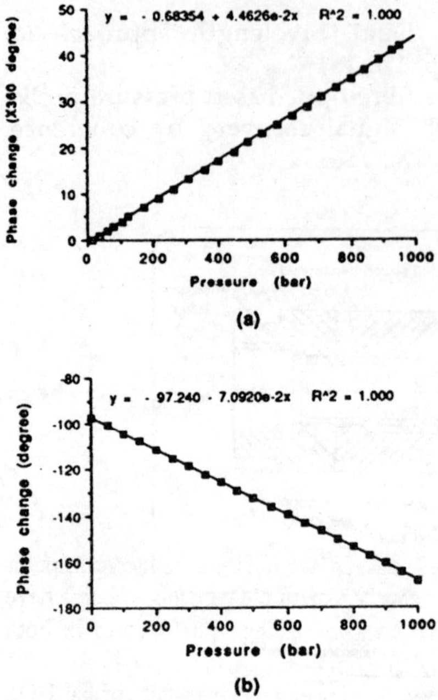


FIG. 2. Response of the pressure probe (a) Variation of $\Phi_{P(\lambda 1)}$ with pressure over a ~1000bar range. (b) Variation of $\Delta\Phi_p$ with pressure over the same range.

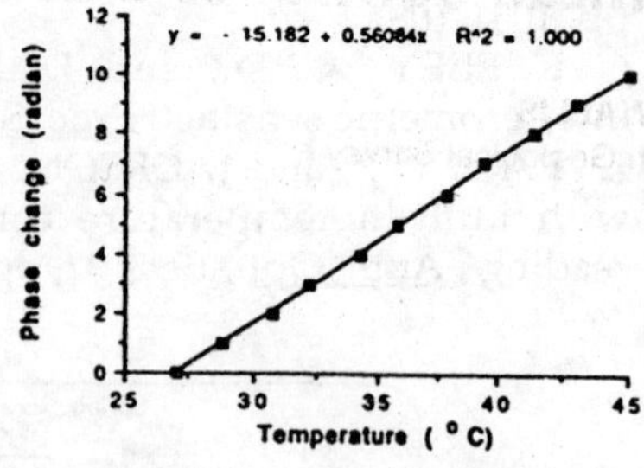


FIG. 3. Response of the temperature probe

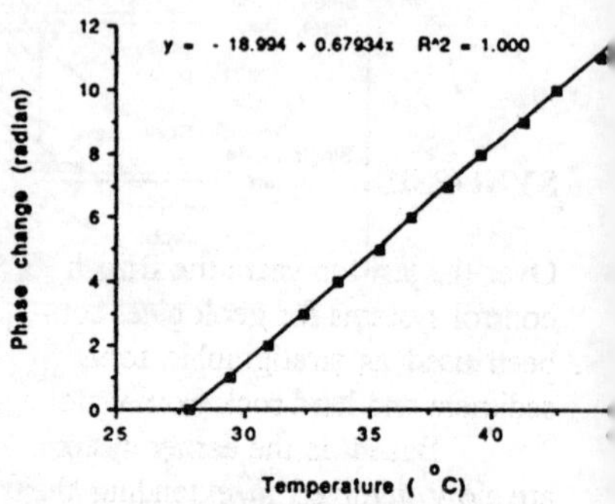


FIG. 4. Thermal response of the pressure probe

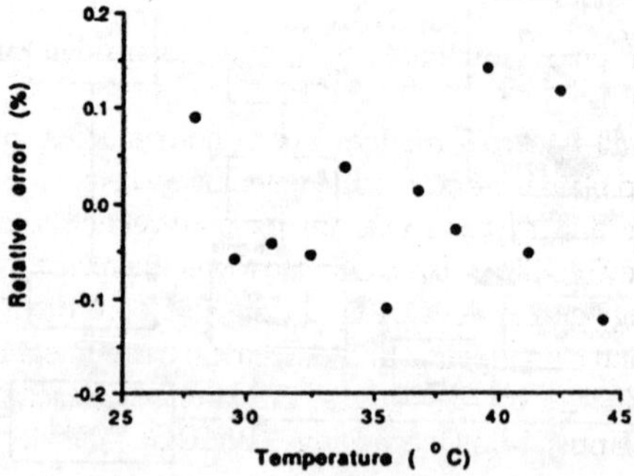


FIG. 5. Relative error distribution after temperature compensation

# emote control of a deep ocean rock drill

G WALLIS
itish Geological Survey

## SYNOPSIS

Over the last ten years the British Geological Survey (BGS) have developed and refined computer control systems for geological coring operations in deep water. The complete coring systems have been used as stratigraphic tools for calibration of seismic data, obtaining seabed cores in both sediment and hard rock from water depths to 2000 metres. (1)

Based on the earlier systems and utilising techniques developed for hard rock coring, BGS are now involved in extending their work into deeper water. This new work will underpin an attempt to recover oriented hard rock cores in water depths to 4000 metres. This paper describes the control elements within the new project work.

## 1 INTRODUCTION

The aim is to retrieve short (50cm) rock cores from mid-ocean ridge environments, oriented to their in situ position.

This new capability will underpin modern ocean floor studies, particularly of mid-ocean ridges, permitting research into plate tectonics using paleomagnetism.

Most of the devices and systems used in its construction are not new, having been developed and refined on previous drills by BGS. However there are considerable problems in extending existing depth ratings and in developing a complete core orientation system which will integrate drilling, orientation and data logging with real time measurement and observation.

The project is being funded by the Natural Environment Research Council (NERC) and comes within the work of the British Mid Ocean Ridge (BRIDGE) initiative.

## 2 SURFACE POWER AND CONTROL

### 2.1 Power supply

The surface power supply will be taken from the research vessel single phase supply and will be controlled through an overcurrent circuit breaker and a residual current circuit breaker for equipment and operator protection. There will be two meters showing the supply voltage and current allowing the operator to monitor the current demands while drilling.

### 2.2 Surface control

Surface control and display is based on an IBM compatible desktop computer (Intel 486/66 is the proposed minimum) running a graphical display programme within a 'windows' environment. An example of the type of display developed for the BGS vibrocorer is shown below (Figure 1).

While drilling, data are logged to hard disc once a second and the display screen is continuously updated. The display shows a graphical presentation of the core barrel penetration into the sea floor plotted against time along with other details such as site reference or location, the date, time and the vessel used. The display includes tabular presentations showing the drill speed, motor currents, external and internal temperature together with heading from the drill mounted sensors.

The display also shows control buttons allowing the operator to send commands to the rockdrill. Selection of seabed functions is achieved using a mouse (or by keystrokes) to maintain real time control during geological coring. Additional monitoring functions from drill mounted sensor data may also be displayed within the surface control system to allow decision making and intervention by the operator during coring operations.

The graphical display described above is written using a commercially available 'C' software package and the desktop computer communicates to the seabed computer using a standard serial interface port at 2400 Baud. This serial data string is converted to a frequency shift keyed (FSK) signal at 180kHz before being sent to the seabed.

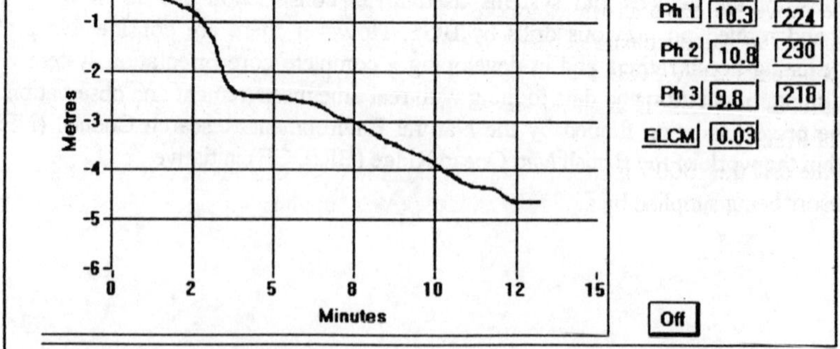


Figure 1 Vibrocorer surface PC control display

## 3 CABLE POWER AND DATA COMMUNICATIONS

### 3.1 Cable selection

As the project is part of the U.K. BRIDGE programme and funds were not allocated to allow purchase of a purpose built cable the drill power and data transmission are being designed into the capacity of existing NERC cables installed on the research vessels supporting the programme..

The NERC Research Vessel Services (RVS) ships supporting the BRIDGE programme are equipped with wire armoured co-axial cables (Table 1). The power and data transmission will be accommodated by these cables which are 10km long.

Table 1 Cable electrical data

| | |
|---|---|
| Voltage Rating | 3000 Volts |
| Insulation: centre-return | 24 000MΩ /m |
| Insulation: return-armour | 1200MΩ /m |
| dc resistance: centre | 4.1Ω /km |
| dc resistance: return | 2.6Ω /km |
| dc resistance: loop | 6.7Ω /km |
| Capacitance @ 1kHz | 108pF /m |
| Attenuation @ 0.5MHz | <4.1dB /km |
| Attenuation @ 1MHz | <5.8dB /km |
| Impedance @ 1MHz | 45Ω |

### 3.2 Mechanical

The cables have a maximum load bearing capacity of 18.4tonnes and a diameter of 17mm. On account of safe working load practices the rockdrill size will be kept below 1tonne. In order to preserve the integrity of the cable, a weak failsafe link will be built into both the drill core barrel and the subsea cable termination so that in the event of the drill becoming fast to the seabed progressive severing of the links will allow recovery of the cable and drill or undamaged cable.

### 3.3 Power and data transmission

The power to drive all the functions on the drill is supplied from the surface using the coaxial cable. The same coaxial cable is to be used for data transmission. It is proposed to transform the vessel 240V 50Hz a.c. supply to 1500V for transmission to the drill. If a maximum power requirement of 4kW is assumed this gives ~15% Voltage drop in the cable using the loop resistance of 67Ω.

At the drill the 1500V is transformed back to 240V for use by the drill. The microprocessor and the sensors being supplied by a number of d.c. power supplies, giving +5V, ±12V and 24V.

The serial communication link will be semi-duplex in that both communications to and from the drill will use the same coaxial cable but there will not be simultaneous transmission. It is proposed to use a frequency shift keying (FSK) system based around 180kHz. This high frequency will be relatively easy to separate from the 1500V power whilst not suffering from excessive attenuation.

## 4 SUBSEA STRUCTURE AND MECHANICS

### 4.1 Structure

The mechanical design of the rockdrill is related to a proven design developed by BGS some years ago. This is based on a triangulated tripod allowing a high strength to weight ratio to accommodate seabed landing forces and the good stability essential for hard rock drilling. The structure is generally within a 2 metre triangular envelope with the power and drive mechanisms located to give as low a centre of gravity as is practical. The electronics pressure case, retract mechanism and power transformer are all located within the structure footprint.

As the corer is intended for use on outcropping hard rock formations no allowance is made for sinking into soft sedimentary overburden.

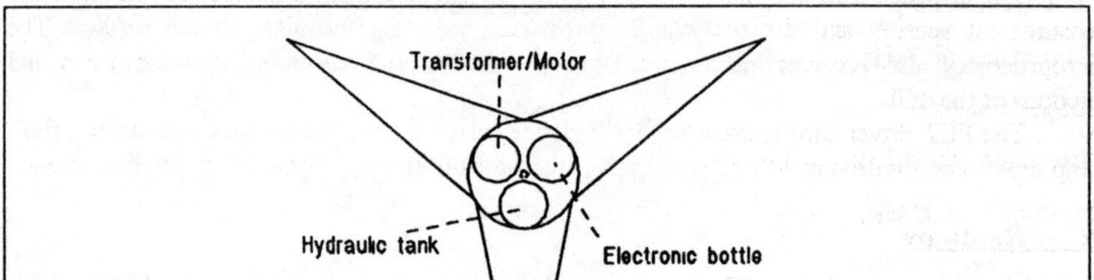


Figure 2 Sketch showing approximate layout of original drill

## 4.2 Motors

There will be three electric motors on the instrument. The drill bit will be turned using a top-drive mechanism driven by a 3 kWatt three phase motor. It is intended to use a proportional speed control for this motor for suitable speed selection dictated by drill bit and rock type.

During drilling a 1kWatt water pump motor will ensure that the drill bit is lubricated by flushing seawater down the annular space between the inner non-rotating core barrel and the external rotating barrel.

Once the drill barrel has penetrated far enough into the sea floor, the drill motor will be switched off and a less powerful, highly geared motor will be switched on to retract the core barrel from the sea bed. A conical spring core-catcher based on a British Standard design will be used to ensure that the rock core is retained by the core barrel as it retracts.

# 5 SUBSEA POWER AND CONTROL

## 5.1 Electronics

These are based round a microprocessor card, an analogue to digital converter (ADC) card, a purpose designed and built interface and signal conditioning (S/C) card and a field effect transistor (FET) relay driver card.

The interface card and the ADC card obtain information from a number of drill mounted measurement sensors and the subsea microprocessor transmits this data to the surface. The microprocessor also receives instructions from the surface and controls the mechanisms and functions of the drill.

The FET driver card is used to switch power relays for the main motor, the drilling flush pump motor and for driving 24V solenoid valves used for the control of the drill hydraulic systems.

## 5.2 Hardware

The subsea processor is an '80188' compatible microprocessor running at 10MHz. This communicates with an ADC card on the STEbus and draws its power from the same STEbus. Both the microprocessor card and the ADC card are commercially available products and were used on the MAST hammer corer project (2). The ADC card collects data from the purpose designed signal conditioning card built by BGS. The microprocessor can send signals to the FET driver card to control the heavy duty relays and solenoids necessary for the operation of the drill motor and flush pump motor.

The signal conditioning card takes the various sensor inputs and converts them to a standard 0-5V analogue signal which may be used by the ADC card. The various inputs include 4-20mA devices, 0-5V or 0-2.5V inputs, pulses from the drill speed counter and a number of other sensor inputs.

The ADC card has 16 analog input channels which allow easy access to most of the sensors through the signal conditioning card. These sensors are read once a second before the information telegram to the surface is assembled. Use is made of an RS485 link on the microprocessor card to interrogate and receive information from the echosounder. Further external expansion of the drill sensors will possibly use this RS485 connection.

The microprocessor controls an analogue output on the ADC card which is used to vary the speed of the drill motor using a modified but commercially available a.c. motor controller.

## 5.3 Software

The controlling firmware programme for the subsea microprocessor is written in C on a desktop personal computer (PC) and compiled using a Borland C++ compiler before being loaded onto erasable programmable read only memory (EPROM) and installed into a socket on the processor card.

The main advantage of the high level language approach is that changes may be made relatively easily at a later stage, such as adding further sensors or changing the communications protocol. The high level language also allows relatively complex algorithms to be developed for processing the seabed sensor data prior to its transmission to the surface.

Use is made, whilst the subsea programme is being developed, of a supervisory firmware programme on the processor card. Using this system a programme is prepared on a PC within the Borland C++ environment before downloading it to the target processor card through a serial port, where it is placed in random access memory (RAM). The supervisory firmware on the processor card then runs the programme allowing the development PC to debug the programme. Once a satisfactory version of the programme is achieved then this may be placed on EPROM and installed in place of the supervisory firmware.

This part of the design draws extensively on techniques developed and experience gained from the MAST hammer corer project (2).

## 5.4 Communication protocol

The subsea electronics package will monitor various rockdrill sensors, such as inclination, pressure(depth), drill speed and system voltages, and make a limited number of decisions based on them before assembling the data into a 'telegram' for transmission to the surface once a second at 2400 Baud. The telegram will consist of ASCII characters with each group of values being separated by a comma(,) and delimited by a carriage return (C/R) and linefeed (L/F). Once the subsea communication has been received by the surface controller a similar telegram will be returned giving commands to the rockdrill. Most of the decisions and commands will be initiated by the drill operator following his interpretation of the data from the rockdrill.

In order to ensure that electrical noise on the cable does not degrade the transmission and reception of these telegrams it is intended to supply a checksum as the last group of ASCII characters in the message. Incorrect reception causing a 'please repeat data' request to be initiated.

## 5.5 Controls

There are a number of controls on the rockdrill some of which are selected automatically by the subsea processor but most of which are controlled by the operator on the surface vessel. Table 2 shows what these are and lists the proposed method of implementation.

The drill motor speed is effected using one of the four analog output channels buffered through the S/C card controlling the motor speed controller.

Drill reverse function is intended to allow unscrewing of the bit and core barrel should they become irretrievably stuck fast in the seabed.

Table 2    List of rockdrill controls

| Function | Control | Method |
|---|---|---|
| Drill Motor ON/OFF | ADC DIO 0 to FET card | 10A Contactor |
| Flush Motor ON/OFF | ADC DIO 1 to FET card | 2A Contactor |
| Retract Motor ON/OFF | ADC DIO 2 to FET card | 5A Contactor |
| Drill Motor Reverse | ADC DIO 3 to FET card | 10 A Contactor |
| Retract Motor Reverse | ADC DIO 5 to FET card | 5A Contactor |
| Drill Motor Speed | ADC DAC 0 to S/C card | 3∅ motor controller |

# 6    SUBSEA SENSORS

## 6.1    Sensors

The drill is equipped with a range of sensors for measuring core barrel penetration, motor speeds, water depth, temperature and so on. Table 3 shows what these are and describes how the information is presented to the signal conditioning card or in the case of the echosounder directly to the microprocessor.

Table 3    Types of sensor used on rockdrill

| SENSOR | SIGNAL | Output | Ref |
|---|---|---|---|
| Altimeter/Echosounder | Drill penetration | RS485 | 6.2 |
| Water Pressure | Water depth | 4-20mA | 6.3 |
| Fluxgate Compass | Heading | 0.1-1.95V | 6.4 |
| Drill rev. counter | r/min | 12V pulses | 6.5 |
| Water Temperature | Temperature | μa/deg Kelvin | 6.6 |
| Subsea Electronics Temperature | Temperature | μa/deg Kelvin | 6.6 |
| Drill Motor Current | Current | 0-2A a.c. | 6.7 |
| Retract Motor Current | Current | 0-2A a.c. | 6.7 |
| Flush Pump Motor Current | Current | 0-2A a.c. | 6.7 |
| Water Conductivity | Conductivity | 4-20mA | 6.8 |
| Altimeter/Echosounder | Drill altitude | RS485 | 6.2 |
| External North-seeking Gyro | Heading | 0-5V | 6.9 |

## 6.2 Drill penetration

This is a high frequency (500kHz) narrow beam echosounder set looking down from the descending drill top drive assembly as it approaches the seabed. The penetration of the drill bit into the seabed is given by subtracting the height of the drill top drive assembly from the overall height of the rockdrill. This echosounder will have a range of some 30metres and a resolution of 1mm. It may be also be used to estimate distance from the seabed as the drill is being lowered from the surface vessel .

penetration = rockdrill height - echosounder/seabed distance

This will be plotted graphically on-screen against time, allowing the operator to monitor penetration rate and will also be logged to disc for use in later drill log analyses.

## 6.3 Water depth sensor

The water pressure sensor is fitted into an endplate of the electronics pressure housing. It is rated to 500bar and has an output of the 4-20mA type. This is used to gauge water depth. The sensor output is converted on the signal conditioning card to fit the range of 0-5V which may be read by the ADC card. The anticipated range of this sensor will be from sea surface to 5000metres depth with an accuracy of 10metres.

## 6.4 Fluxgate compass

This has a variety of outputs and at present the analogue voltage output is being utilised. This is passed through the signal conditioning card before being used by the ADC card. It may become more accurate or sensible to use the serial data string from the compass input directly to the processor card. This option is yet to be explored.

The compass depends on a surface calibration during which the rockdrill is rotated through 360° at least 100 metres from the ship, this calibration makes an automatic allowance for the ferrous material on the rockdrill. The calibration facility is part of the software integral to the fluxgate compass sensor unit.

Although the local magnetic field will affect the compass, the extent of the effect is uncertain. Similar basaltic lavas in Hawaii generate a magnetic field strength of 10A/m at 0.5metres altitude (3). This should give directional errors of less than 5° using the fluxgate compass. In an attempt to minimise the local effects the compass will be mounted in an external pressure housing as far from the seabed as is possible on the rockdrill (~ 1.5metres).

## 6.5 Drill revolution counter

At the top of the drill core barrel a bronze bearing assembly is designed into the mechanism 25cm in diameter with 50 small cylindrical magnets buried in its circumference. Magnets are inserted so that each magnet has opposite polarity to its two neighbours. A magnetic flux detector positioned adjacent to the rim of the wheel produces a count as the wheel turns. The series of pulses created, one pulse for each magnet, when applied to an automotive tachometer integrated circuit on the signal conditioning card, will generate a 0-5V signal for use by the ADC card.

The effect of the magnetic field from this sensor should be effectively removed from the heading information by the calibration facility within the fluxgate compass.

The processor card will either automatically regulate the drill speed to a pre-determined value or may simply relay the actual drill speed to the surface control operator for his evaluation and decision.

## 6.6 Temperature sensors

These are based on sensor assemblies which have an output current proportional to the temperature. These sensors will be calibrated at 25° Celsius to minimise the errors which may arise. As these sensors are susceptible to pressure the external sensor is placed within a small pressure housing outside the main electronics pressure bottle. The main internal sensor is mounted on the signal conditioning card with provision to site another temperature sensor if necessary within the drill power transmission system.

Although the water temperature at the target sites can exceed 100° C the electronics assemblies are not intended for operation above 70° C and higher external temperatures may necessitate selecting an alternative sample site.

## 6.7 Motor currents

The drill drive motor, flush pump motor and retract motor currents will be measured using current transformers. The signal conditioning card will convert the a.c. current to a d.c. voltage for use by the ADC card.

Both the drill and flush pump motors will be supplied through electrically resetable circuit breakers. These will detect excess current and protect the motors. Earth leakage currents will be monitored at the surface.

## 6.8 Water conductivity sensor

The water conductivity sensor, whilst not of primary importance, is there to help give an overall picture of the environment on the seabed. As with the pressure sensor it has its output converted to fit within the scope of the ADC card. It will be mounted as part of the end plate assembly of the electronics pressure housing.

## 6.9 External North seeking gyrocompass

This input is to allow the inclusion on the package of a North seeking gyrocompass to verify the accuracy and precision of the fluxgate unit which will be permanently mounted on the rockdrill. The subsea electronics will be capable of either reading a serial data string and/or interpreting an analogue voltage before relaying the reading to the surface. This item will probably only be used on an occasional basis.

This sensor will require a 4 pin connection from the electronics pressure container to the external part of the drill. This external connection may well be used for other ancillary equipment as the project develops.

# 7 CONCLUSIONS

The project is currently at the desk study evaluation and prototype design stage. It is scheduled to run for some two years of which 6 months have passed. It is likely that some of the ideas outlined within this paper may be implemented in alternative ways as the rockdrill develops.

Prior to planned prototype sea trials and system function tests the development will be subjected to land based tank trials, as with previous work.

Once the development is finished, it should furnish a useful tool for examining the seabed outcrops, particularly on the mid ocean ridges. It may need supplementary data from other sensors and instruments in order to help provide a clearer picture of the environment at those depths. It should complement other development work currently occurring within the BRIDGE iniative.

## 8 ACKNOWLEDGEMENTS

The Author of this paper acknowledges the contributions and assistance of his colleagues in creating this research tool, in particular J.B.Pheasant and D.J.Smith of BGS and Dr S. Allerton of the Grant Institute, Edinburgh University.

Published with the permission of the Director of the British Geological Survey.



**References**

1 PHEASANT, J.B. A Microprocessor Controlled Seabed Rockdrill/Vibrocorer, SUT, Underwater Technology, 1984, 10, 10-14.

2 SMITH, D.J. and TRONSTAD, K. Deep Sea Hammer Corer, Electronic Engineering in Oceanography, Churchill College Cambridge, 1994, 129-134, IEE

3 BAAG, C., HELSEY, C.E., SHI-ZHE XU and LIENERT, B.R. Deflection of paleomagnetic directions due to magnetization of the underlying terrain, Journal of Geophysical Research, volume 100, No. B7, pages 10 013 - 10 027.